"[*The Revenge of Analog*] is at its best when highlighting the surprising ways that the digital world has breathed new life into analogue forms."

—*Financial Times*

"No matter which side you're on in the debate over digital technology, there's something to cheer you in *The Revenge of Analog*." —*New York Times Book Review*

"Here is a compulsively readable book after a Luddite's heart.... Sax isn't preaching a return to the pre–industrial age, but neither is he embracing the robot overlords. He thoughtfully, wisely, and honestly points out how analog experiences enhance digital creativity and how humans benefit from what both have to offer. Essential reading." —*Booklist*, starred review

"Beguiling.... I defy you to read Sax's book without wanting to buy a Moleskine, put an LP record on the turntable, or play a game of Scrabble with your friends."

—Bill McKibben, *New York Review of Books*

"A thoughtful look at the many ways in which analog has not been eliminated from the world but…is still thriving.... Sax's book reminds us that we live in an analog world. It is a good reminder that digital can only take us so far."

—Yahoo! Finance

"We all thought the digital age would be the end of analog media—and we were wrong. In this smart, funny, glorious book, David Sax explains why so many of us still crave the tactile, sensual experience of listening to music on vinyl records and taking notes with pencil and paper. Turn off your electronic devices, find a quiet place, and savor this remarkable book." —Dan Lyons, bestselling author of *Disrupted*

"Hang on digital mavens, the real world ain't going anywhere. In *The Revenge of Analog*, David Sax shows the continued importance of the physical stuff to how we live and work today." **—Richard Florida, author of** ***Rise of the Creative Class***

"Sax's lively, evocative prose conjures reminders of the physical world: record presses spit and heave, cameras satisfyingly click, and paper crinkles and smells in ways pleasingly familiar." **—*Globe and Mail* (Canada)**

"The better digital gets, the more important analog becomes. In this fun tour of modern culture, David Sax has collected hundreds of ways that an analog approach can improve our newest inventions. Sax's reporting is eye-opening and mind-changing." **—Kevin Kelly, founding executive editor of** ***Wired* and author of *The Inevitable***

"David Sax has written a brilliant *cri de coeur* about the way things used to be, should be, and, increasingly, are becoming once again. *The Revenge of Analog* reminds us that it wasn't so long ago that records were vinyl, laces were double-knotted, and the mailbox at the end of the driveway was lovingly banged up. It's a book that brings something even more rare than a perfect song at the perfect moment—hope." **—Rich Cohen, cocreator of HBO's *Vinyl* and author of *The Sun & The Moon & The Rolling Stones***

"The more advanced our digital technologies, the more we come to realize that reality rules. David Sax reassures us surviving members of team human that material existence is alive and well, and makes a compelling case for the reclamation of terra firma and all that comes with it." **—Douglas Rushkoff, author of *Throwing Rocks at the Google Bus***

Praise for *The Revenge of Analog*

"[*The Revenge of Analog*] is at its best when highlighting the surprising ways that the digital world has breathed new life into analogue forms."

—*Financial Times*

"No matter which side you're on in the debate over digital technology, there's something to cheer you in *The Revenge of Analog*."

—*New York Times Book Review*

"Here is a compulsively readable book after a Luddite's heart. . . . Sax isn't preaching a return to the pre-industrial age, but neither is he embracing the robot overlords. He thoughtfully, wisely, and honestly points out how analog experiences enhance digital creativity and how humans benefit from what both have to offer. Essential reading."

—*Booklist*, starred review

"Beguiling. . . . I defy you to read Sax's book without wanting to buy a Moleskine, put an LP record on a turntable, or play a game of Scrabble with your friends. . . . At the outset of this review I compared the digital era to a fast-moving stream, which theoretically one could learn to navigate. But it's more likely, I think, that we're in a permanent flood stage, where we will have to somehow continue stretching and contorting ourselves to stay above the tide or else resign ourselves to drowning in the cascade of data. One is grateful to David Sax for mapping the eddies where we might, at least for a moment, find some stillness, respite, and fun."

—Bill McKibben, *New York Review of Books*

"A thoughtful look at the many ways in which analog has not been eliminated from the world but . . . is still thriving. . . . Sax's book reminds us that we live in an analog world. It is a good reminder that digital can only take us so far."

—*Yahoo! Finance*

"Sax's lively, evocative prose conjures reminders of the physical world: record presses spit and heave, cameras satisfyingly click, and paper crinkles and smells in ways pleasingly familiar."

—*Globe and Mail* (Canada)

"Hang on digital mavens, the real world ain't going anywhere. In *The Revenge of Analog*, David Sax shows the continued importance of the physical stuff to how we live and work today."

—Richard Florida, author of *Rise of the Creative Class*

"We all thought the digital age would be the end of analog media—and we were wrong. In this smart, funny, glorious book, David Sax explains why so many of us still crave the tactile, sensual experience of listening to music on vinyl records and taking notes with pencil and paper. Turn off your electronic devices, find a quiet place, and savor this remarkable book."

—Dan Lyons, bestselling author of *Disrupted*

"The better digital gets, the more important analog becomes. In this fun tour of modern culture, David Sax has collected hundreds of ways that an analog approach can improve our newest inventions. Sax's reporting is eye-opening and mind-changing."

—Kevin Kelly, founding executive editor of
Wired and author of *The Inevitable*

"David Sax has written a brilliant *cri de coeur* about the way things used to be, should be, and, increasingly, are becoming once again. *The Revenge of Analog* reminds us that it wasn't so long ago that records were vinyl, laces were double-knotted, and the mailbox at the end of the driveway was lovingly banged up. It's a book that brings something even more rare than a perfect song at the perfect moment—hope."

—Rich Cohen, cocreator of HBO's *Vinyl* and
author of *The Sun & The Moon & The Rolling Stones*

"The more advanced our digital technologies, the more we come to realize that reality rules. David Sax reassures us surviving members of team human that material existence is alive and well, and makes a compelling case for the reclamation of terra firma and all that comes with it."

—Douglas Rushkoff, author of *Throwing Rocks at the Google Bus*

THE
REVENGE
of
ANALOG
REAL THINGS *and*
WHY THEY MATTER

DAVID SAX

PUBLICAFFAIRS
NEW YORK

PublicAffairs
Hachette Book Group
1290 Avenue of the Americas, New York, NY 10104
www.publicaffairsbooks.com
@Public_Affairs

Printed in the United States of America

First Trade Paperback Edition: October 2017
Published by PublicAffairs™, an imprint of Perseus Books, LLC,
a subsidiary of Hachette Book Group, Inc.

The Hachette Speakers Bureau provides a wide range of authors for speaking events.
To find out more, go to www.hachettespeakersbureau.com or call (866) 376-6591.
The publisher is not responsible for websites (or their content) that are not owned by the publisher.

Print Book Interior Design by Jack Lenzo

The Library of Congress has cataloged the hardcover edition as follows:
Names: Sax, David, author.
Title: The revenge of analog : real things and why they matter / David Sax.
Description: First Edition. | New York : PublicAffairs, 2016. | Includes
    bibliographical references and index.
Identifiers: LCCN 2016012413 (print) | LCCN 2016027805 (ebook) | ISBN
    9781610395717 (hardback) | ISBN 9781610395724 (ebook)
Subjects: LCSH: Entrepreneurship—History. | Electronic commerce—History. |
    Marketing. | BISAC: BUSINESS & ECONOMICS / Industries / Retailing. |
    SOCIAL SCIENCE / Popular Culture. | TECHNOLOGY & ENGINEERING / Manufacturing.
Classification: LCC HB615 .S3137 2016 (print) | LCC HB615 (ebook) | DDC 306.3—dc23
LC record available at https://lccn.loc.gov/2016012413

ISBNs: 978-1-61039-571-7 (hardcover); 978-1-61039-572-4 (ebook); 978-1-61039-821-3 (paperback)

LSC-C

10 9 8 7 6 5 4 3 2

Contents

A new medium is never an addition to an old one, nor does it leave the old one in peace. It never ceases to oppress the older media until it finds new shapes and positions for them.

—Marshall McLuhan, 1964

JACKIE TREEHORN: "New technology permits us to do very exciting things in interactive and erotic software. The wave of the future Dude, one hundred percent electronic."
THE DUDE: "Hmm . . . well, I still jerk off manually."

—*The Big Lebowski*, 1998

Introduction

*I*n *June 2012, a new* store called June Records opened its doors in Toronto's Little Italy neighborhood, a block and a half from a house I'd just purchased with my wife. June Records looked nothing like the dusty, cluttered record stores I'd grown up visiting. It was a modern, well-organized, well-lit retail space, almost a boutique. Walking through the neighborhood shortly after I bought the house, I was stopped on the sidewalk outside June Records by the beautiful sounds coming from the turntable playing in the window. The album was Aretha Franklin's *Live at the Fillmore West*, and the combination of the Queen of Soul and a sunny summer day in my new neighborhood was impossible to resist. I walked in, asked how much the record cost, and walked away with Aretha under my arm, practically dancing down the sidewalk with elation.

Like most music fans, I'd spent the preceding decade gradually divorcing my music collection from physical reality: loading CDs into iTunes, iTunes onto my iPhone, and eventually the whole deal into the cloud. I still owned a turntable, an old Technics given to me by my friend David Levy, but it had been sitting in a box at my parents' house for more than two years, utterly neglected. The $20 I had spent on that record at June was more money than I had spent on music of any kind over the same time.

That fall, we moved into our house, I set up my turntable, and finally played Aretha in all her sweet glory. Within a few bars of "Respect," it dawned on me that this was the first recorded music I had actively listened to in ages. It had been months since I had opened iTunes on my computer, and I had no songs anymore on my phone.

All those albums were hidden in my hard drive, nestled between old e-mails and various other files, beyond my sight. Most days I just listened to public radio in the kitchen or the car. When my brother bought me a subscription to the streaming service Rdio, I frequently found myself opening up the app, only to become paralyzed with indecision. My options were infinite, literally every single album and song ever recorded. What did I want to listen to? It was as though the ease and convenience of digital music had sucked the very fun out of listening to it. The entire world of music was just a click away, but I couldn't even be bothered to do that. What if there was an even better song, just a few taps away? Something was missing. The way to bring it back, I now realized, was vinyl.

I found myself purchasing records whenever I could. I began by digging through the discount bins at June Records for old jazz and soul albums, but was very soon buying freshly pressed, newly released records by bands and artists I learned about by actually talking to June's staff. Often, if something was playing on the turntable there, I'd pick it up: the harmonic guitar rock debut from local band Always, a fresh compilation from Nigerian psych-disco hermit William Onyeabor, the stripped-back instrumental hip-hop collaboration from rap legend Ghostface Killah and funk band BadBadNotGood. My new collection, which numbered maybe a dozen old records when I moved into the house, swelled so quickly, my wife declared a hard limit on the number of shelves it could occupy.

I was having too much fun to care. My modest vinyl fetish had tapped into something that was lying dormant since my first Napster download: the carnal pleasure of physically browsing and buying music. I would walk by a record store and suddenly the $10 bill in my wallet would begin burning, demanding to be spent. Half an hour later I would emerge with an album under my arm, my face swollen with pride, as though I had recorded the damn thing myself. In an age when I could have the exact same music for free, and play it on five different devices, here I was paying good money for scratchy, heavy, cumbersome discs of melted plastic that I had to play on a machine as temperamental and costly to maintain as an old car. It was totally irrational.

I wasn't alone in this madness. Every few months, I'd stumble on a new record shop that had just opened, or an older one that was suddenly expanding to a second or third location. Each time one did, it seemed like a minor miracle. Record shops had been doomed to extinction in the collective imagination ten years before, deployed as a metaphor for dying retail businesses that failed to adapt to the digital era ("If bookstores don't figure out the web, they could go the way of the record store"). Nobody opened new record stores. No one.

Then, overnight it seemed, these retail dinosaurs were not only back from the brink but walking among us, and multiplying in every corner of the world. News stories about the death of the record store were replaced by stories about the anomaly of record stores defying the odds, new record stores like June opening, and finally confident declarations that record stores were not only back, but were actually thriving. The number of new vinyl records pressed and sold has increased more than tenfold over the past decade, resulting in a similar boom in turntable sales, and record stores opening. June Records, for example, has grown its sales by roughly 5 percent, every single month since it opened, according to the store's co-owner Ian Cheung, doubling its revenue pretty much every year that June Records has been in business. Just last month, yet another record shop opened a few blocks away from June. Cheung told me he isn't worried about competition. The more record stores that are out there, the more relevant June seems.

More significant than the sales was the demographic buying these records. Andrew Zukerman, a long-haired, appropriately opinionated clerk at June (basically your prototypical record store guru) characterized the average customer of record stores over the previous decade as "crummy old men looking for great records in dollar bins." You know the type: graying ponytail secured behind a balding crown, tattered Jane's Addiction concert T-shirt tucked into patched black jeans, endless soliloquies of cultural superiority emerging from their lips.

Around the time June Records opened, something dramatic changed in its customer base. The crummy geezers were very quickly supplanted by younger customers; music lovers in their twenties and even teens, kids who grew up with digital music, people who had *only* ever listened to music as free virtual files on Apple devices. And there was one other demographic that came as a surprise.

"Girls!" said Cheung, with the exasperated relief of a man who stumbles upon a river after wandering the desert. "When girls started buying records once again, you knew things had changed." Zukerman just nodded his head. "When girls started shopping for vinyl, you saw the look in these old guys' eyes." It was a look of fear. The return of the female vinyl record buyer, in ever growing numbers, signaled the return of the record shop to its proper place in the cultural landscape; somewhere young people came to discover music, and one another. Somewhere cool.

Everyone from the media to the music industry struggled to explain the surprising return of vinyl records worldwide. Commonly cited marketing buzzwords, such as *authenticity, nostalgia*, and *millennial*, were deployed in various combinations. Others just pinned it on the dreaded *hipster*, that ill-defined species of early aughts youth culture, which remains the preferred scapegoat for any urban gripe, from gentrification to the tightness of jeans.

I saw the return of vinyl records as part of a bigger phenomenon. The Revenge of Analog.

———•———

*F*ive years before June Records opened, I attended a retreat in Park City, Utah, organized by a Jewish organization called Reboot. The weekend included all sorts of activities designed to reexamine Jewish identity and culture, and a part of this required everyone in attendance to abstain from technology over the Sabbath, the twenty-four-hour rest period between sundown Friday and sundown Saturday. When I unplugged, I found the experience so restorative that I regularly began observing my own digital Sabbath when I returned to Toronto, even though I wasn't religious in any other way.

A few weeks later, my girlfriend (now wife) and I were invited to a Friday night dinner at a friend's place. There were eight of us at the table, and during the entire meal, everyone except my wife and I held a BlackBerry smartphone, typing away through appetizers, mains, and dessert. We just sat there dumbstruck, kicking each other under the table each time someone blanked on the conversation and returned to their tiny keyboard, while their chicken grew cold. This was the first time we had both witnessed such a fundamental change in human social behavior at technology's hands, and it shook us both to the core.

Of course, we were observing just the tip of the iceberg looming on the horizon. Within months of that night the first iPhone would appear, and all of us would embrace its seemingly limitless capability to captivate our attention. Soon, my wife and I were just like every other couple; our faces buried in screens at the dinner table, blind to the world around us, and to each other.

Later that night, back at the apartment I shared with my good friend Adam Caplan, the story of the awkward dinner quickly turned into a long conversation about the way digital technology was changing our lives. Adam, who is a teacher and extremely tech savvy, strongly believed in the transformational power of digital technology. But he openly acknowledged that digital's gain was not without sacrifice.

Adam had recently brought over a turntable from his parents' house, along with the bulk of their records (including the complete works of Herb Alpert and the Tijuana Brass), and that vinyl collection provided the soundtrack to, and impetus for, that pivotal chat.

The experience of listening to a record was less efficient, more cumbersome, and not necessarily sonically superior to a digital file played on the same stereo. But the act of playing a record seemed more involved, and ultimately more rewarding, than listening to the same music off a hard drive: the physical browsing of album spines on the shelf, the careful examination of the art on the sleeve, the diligent needle drop, and that one-second pause between its contact with the record's vinyl surface and the first scratchy waves of sound emerging from the speakers. It all involved more of our physical senses, requiring the use of our hands, feet, eyes, ears, and even mouth, as we blew dust from the record's surface. There was a richness to the vinyl record experience that transcended any quantifiable measurement. It was more fun precisely because it was less efficient.

That experience was different, Adam told me, because it was analog. Analog, in the broadest sense of the term (and the one I've grounded this book in), is the opposite of digital. Digital is the language of computers, the binary code of 1's and 0's, which in endless combinations allow computer hardware and software to communicate and calculate. If something is connected to the Internet, runs with the help of software, or is accessed by a computer, it is digital. Analog is the yin to digital's yang, the day to its night. It doesn't require a computer to function, and most often analog exists in the physical world (as opposed to the virtual one).

As I began to see the world through this prism, I noticed something going on. Certain technologies and processes that had recently been rendered "obsolete" suddenly began to show new life, even as the world around them was increasingly driven by digital technology. Every week I'd walk down the street and find a new boutique focused on an analog pursuit that had nothing to do with computers: letterpress cards and invitations, film photography, handmade leather goods and watches, new print magazines, fountain pens, and, of course, vinyl records. A board game café opened up around the corner from our apartment, and it had a line out the door from day one.

I spend most of life as a writer, but I also do some investing in startups, and the trend I was noticing ran counter to the standard narrative around innovation in our economy. Everyone was supposed to be on the cusp of inventing the next great app, but the new businesses that seemed to matter in my life were something else entirely. They were places with walls and windows that sold things you could hold in your hand.

It was as though analog was becoming newly relevant, right as its very obsolescence was supposedly assured. The Revenge of Analog represented a resurgent and reimagined value for nondigital goods, services, and ideas, precisely when the transition from analog to digital was supposed to be total. But as digital technology assumed an increasingly large role in our lives, it almost seemed as if an alternative, postdigital economy was emerging as well. Although I first saw this trend blossoming in trendy urban neighborhoods, it quickly spread out into mainstream consumer culture. Initially it appeared as a trickle, but soon the Revenge of Analog was a torrent rushing all around me.

I'd meet with a technology company founder at a suburban Starbucks, and he'd be taking notes in a Moleskine journal, along with every other person in the place. A chain store, such as Urban Outfitters, suddenly carried a vast selection of new Polaroid film cameras, while Whole Foods announced it would be selling vinyl records. Each day the news would feature a new story about a fresh analog trend. Courses on mindful meditation, and luxury corporate retreats that forced participants to unplug, were proliferating in Silicon Valley. Books about the perils of digital distraction and the benefits of face-to-face interactions became best sellers. The very same friends who ignored me during that dinner a decade back now put their phone away during meals. Academic studies about the importance of real-life interactions, screen-free parenting, and reading on paper appeared regularly. Amazon actually opened a physical bookstore in Seattle, following other online retailers into the brick-and-mortar locations they once vowed to topple. Even cassette tapes reappeared. All of a sudden, *analog* was a buzzword.

How had this happened? Had I drifted into some sort of Wes Anderson–directed fever dream of handmade, curated preciousness? Was I simply noticing all of this because it fascinated me, or was something deeper driving this? Had our digital love affair reached some threshold, triggering a shift in momentum away from the tremendous and seemingly inevitable march of digital progress that had defined my entire life, for reasons that spoke to a deeper human truth? In a world increasingly defined by digital technology, why was I witnessing the Revenge of Analog?

———◆———

*E*very day we turn around and something else has been enhanced, altered, or shaken up by digital technology: our car, our house, our job, our sex life. In the clean, orderly narrative of technological progress, the newest technology always renders the old one obsolete. We evolved from listening to music live, to listening on wax cylinders, then vinyl records, cassette tapes, compact discs, MP3 downloads, and now wireless streaming services. The future of music listening clearly points to services that are cheaper, faster, higher quality, less costly, and entirely virtual, as it does for so many other elements of our lives touched by digital technology.

Up until very recently, if something could be digitized, its fate was a foregone conclusion. Magazines would only exist online, every purchase would be made through the web, classrooms would be virtual. Any job that could be performed by a computer was already redundant. Our world would be successively rendered into bits and bytes, one program at a time, until we reached a state of digital utopia, or the Terminators came for us.

The Revenge of Analog presents a different narrative, however. It shows that the process of technological innovation isn't a story of a slow march from good to better to best; it's a series of trials that helps us understand who we are and how we operate.

The Revenge of Analog is occurring now precisely *because* digital technology has become so damn good. Digital computing has been with us for the better part of the past half century, personal computing for the past three decades, the Internet for two decades, and smartphones for one. Today, a digital solution is almost always the default: the most efficient, widely used, cheapest, and obvious tool to get the job done. With a few finger taps you can just as easily order a warm cookie to your house as set up a massive data center in the cloud.

Because of this, digital's overwhelming superiority initially renders the analog alternative largely worthless, and devalues that analog technology significantly. But over time, that perception of value shifts. The honeymoon with a particular digital technology inevitably ends, and when it does, we are more readily able to judge its true merits and shortcomings. In many cases, an older analog tool or approach simply works better. Its inherent inefficiency grows coveted; its weakness becomes a renewed strength.

This is why the Revenge of Analog matters, and why the rising value of analog goods and ideas I write about in this book is just the beginning. Surrounded by digital, we now crave experiences that are more tactile and human-centric. We want to interact with goods and services with all our senses, and many of us are willing to pay a premium to do so, even if it is more cumbersome and costly than its digital equivalent.

The Revenge of Analog not only questions our assumptions about digital's inevitability, but the very certainty at the heart of the digital economy. This is a powerful current to swim against. The notion that nondigital goods and ideas have become more valuable would seem to cut against the narrative of disruption-worshipping techno-utopianism coming out of Silicon Valley and other startup hubs, but, in fact, it simply shows that technological evolution isn't absolute. We may eagerly adopt new solutions, but, in the long run, these endure only if they truly provide us with a better experience—if they can compete with digital technology on a cold, rational level.

This is where the Revenge of Analog matters even more. While analog experiences can provide us with the kind of real-world pleasures and rewards digital ones cannot, sometimes analog simply outperforms digital as the best solution. When it comes to the free flow of ideas, the pen remains mightier than both the keyboard and touchscreen. And as you'll see throughout this book, the natural constraints analog technology imposes on its users can actually increase productivity, rather than hinder it.

Let me be clear that this book is not a screed against digital technology. The individuals, companies, and organizations you're going to meet here are not driven, in any way, by rose-colored nostalgia for an idealized, predigital past. There is not a Luddite among them. They are incredibly forward thinking and innovative, and use every digital tool at their disposal—online crowdfunding, social media, design software, smartphones—to bring analog goods and services to market. They aren't pushing the digital world away. Rather, they're pulling the analog one closer, and using its every advantage to succeed.

Because of this, the Revenge of Analog represents a tremendous opportunity for the companies and institutions embracing the universal truth behind this sweeping phenomenon. There is money to be made addressing the desire of consumers for new and reimagined analog products and services, but there is also a deeper lesson we can all learn about the way we interact with the world, and how our choice of technology shapes that. Even if you never plan to listen to a vinyl record, play a board game, or manufacture watches in Detroit, the lessons from those who have succeeded in these businesses are undeniably valuable to anyone, or any organization, looking to live, and thrive, in our postdigital economy.

I have written this book in two halves that explore how this occurs:

Part I: The Revenge of Analog Things examines the new markets for vinyl records, paper products, film photography, and board games, to show how the businesses that make and sell legacy analog goods succeed today by tapping into the fundamental consumer desires driving the growth of these products.

Part II: The Revenge of Analog Ideas draws lessons from publishing, retailing, manufacturing, education, and even Silicon Valley, to demonstrate the innovative and disruptive potential of analog ideas in today's digitally focused economy, and the advantages they bring to those who embrace them.

The choice we face isn't between digital and analog. That simplistic duality is actually the language that digital has conditioned us to: a false binary choice between 1 and 0, black and white, Samsung and Apple. The real world isn't black or white. It is not even gray. Reality is multicolored, infinitely textured, and emotionally layered. It smells funky and tastes weird, and revels in human imperfection. The best ideas emerge from that complexity, which remains beyond the capability of digital technology to fully appreciate. The real world matters, now more than ever.

The Revenge of Analog is a product of that messy reality. It faces the challenges presented by digital technology and actually draws strength from them. Each technology feeds a different purpose, and produces a different result. What the Revenge of Analog shows us is a model for an emerging postdigital economy that looks toward the future of technology, without forgetting its past.

Before we go further, I want to ask a favor of you, dear reader. This is a book, and though it was written on a computer, and you might be reading it on an electronic device, it works best in an analog context. So, please turn off your phone. Shut out the digital world as best you can, and embrace the silence while you turn the page. Have a seat someplace comfortable, and if you have a turntable, turn it on. We are going to start our journey through the Revenge of Analog the same way I did, through records, and a visit to the heartland of the global vinyl boom.

PART I

THE REVENGE OF ANALOG THINGS

1

The Revenge of Vinyl

The factory floor at Nashville's United Record Pressing (URP) is a living, breathing dragon of place. In the relatively small, windowless warren of concrete rooms, twenty-two record presses hiss, cough, growl, grind, and ultimately spit out vinyl records of all genres, weights, colors, and sizes. A Dave Matthews Band album is born right next to classic reissues from Primus, Pearl Jam, and the Wu-Tang Clan; chart toppers by Lana Del Rey; colored Iron Maiden collectors' editions; a special release by Elvira "Mistress of the Dark"; and neo disco by Chromeo. The place smells like hot metal, sour water, and the sweet poison tang of warm plastic.

Dozens of workers hover around the machines, feeding them steam, water, grease, power, and black pellets of polyvinyl chloride (PVC, a.k.a. vinyl). They gather the records that spit out onto long metal stakes, to make room for more that keep piling up. The machines—great bulky assemblages of hydraulics, heavy-gauge buttons, pipes, hoses, and thick slabs of metal made decades ago—collectively cast off a drone so loud that it's as though they're not only working to mash the impressions of sound waves into hot pucks of melted vinyl, but somehow emitting a primal scream of all the music buried up inside this place, the source of an analog revival so relentless, it has pushed these old record presses to their breaking point.

Visit United Record Pressing as recently as 2010 and things would have been much quieter. A lot of the time the presses would be turned off, waiting for a new order, and two thirds of the people scurrying about today would have been working elsewhere. Back then, United Record Pressing was at its low point: work had dropped to one six-hour shift, maybe fifty people, just a couple of days a week. The owners had to rely on loans to keep the place open. On an average day URP was pressing a few thousand records, and those numbers kept declining, as they had for most vinyl record–pressing plants worldwide since the early 1990s.

Four years later, as I stood between those machines and their gorgeous cacophony, they were pressing forty thousand records every single day, operated by a staff that had tripled in size since 2010. They worked twenty-four hours a day, six days a week, with only the Lord's Day for rest. Today, URP's orders pile up like the stacks of freshly pressed albums on its shipping dock: it takes two to three months of waiting to get your record pressed if you are a big label, many more for independents. Until recently the company was so swamped, it refused any new customers. It can barely get the records out the doors fast enough, which is why United Record Pressing must grow, so it can press more records. Music fans are hungry. Their appetite for vinyl is voracious, and there's no sign they are going on a diet anytime soon.

Wherever there is a middle class of music lovers, you will find a significant and growing percentage of them buying turntables and records to play on them. These include old records dug up from basements, vintage records purchased online and in stores, and increasingly, new records pressed daily at such plants as United Record Pressing. One European plant owner estimated nearly 30 million new records were pressed worldwide in 2015.

Nowhere has vinyl's rediscovery been as widespread, or dramatic, as the United States of America, where United Record Pressing is the nation's largest record-pressing plant, and one of the top three in the world (Germany's Optimal and GZ in the Czech Republic are somewhat larger). From URP's low point in 2010, it has seen business grow so dramatically that the company announced in mid-2014 it would open a second factory nearby, increasing its tally of pressing machines from twenty-two to thirty-eight, and its staff from 150 to more than 250. Here was a clear front line for the vinyl record's resurgence I had experienced firsthand at June Records, and with it, the postdigital economy that was growing to meet a renewed demand for analog goods.

There's a reason they call Nashville Music City. You can't swing a locally made Gibson guitar without hitting someone or something associated with the music business. From the Grand Ole Opry and the Johnny Cash Museum to countless recording studios and country bands playing on the Broadway honky-tonk strip, music seems to be Nashville's driving force. The sweet country whine of a slide guitar still typifies the Nashville sound, but recently, Nashville has seen an influx of rock and indie musicians move to town, drawn by cheap rent, ample studio space, and a deep community of talent. Today's Nashville sound is as closely associated with the raw roots rock of Jack White and the Black Keys, along with Taylor Swift's powerful pop, as it is with fiddles and songs about pickup trucks.

Located just south of downtown, on an industrial strip of warehouses and factories in Wedgewood Hill, the entrance to United Record Pressing features two oversize records embedded in its facade. As you walk through the parking lot, little melted pieces of vinyl crunch under your feet. Inside, everything seems to be made of vinyl: the records framed on the walls and stacked on floors, the midcentury chairs, lamps, desks, floor tiles, and wood-grain wall panels. Aside from paper album covers, metal pressing machines, yellowing photos of such artists as Lionel Ritchie and Rick James, and the humans who work there, pretty much everything in the place is some form of melted petroleum by-product pressed into the service of music.

The company began in 1947 as Bullet Plastics, and was the first record-pressing plant in town. A few years later, Bullet changed its name to Southern Plastics, and eventually to United Record Pressing. The current building has been occupied since 1962, and has manufactured the records at the core of twentieth-century popular music: singles from Elvis and Johnny Cash on Sun Records, the heyday of

Motown and Stax, and even the first Beatles record pressed in America. If it spun on a turntable in America, odds are pretty good that it was pressed in this building.

The place is a time warp. Upstairs there's an apartment with furniture that hasn't changed since the Kennedy administration. It's called the "Motown Suite," because black music executives could stay there when Nashville was segregated. It features a bedroom with a pair of black leather shoes on the floor, which have remained in place for decades, because no one is sure whether they belonged to someone notable, such as Smokey Robinson, or some random schmo who forgot his shoes.

"Music is just vibrations in the air," said Jay Millar, the director of marketing at URP at the time (he now works at the label Sundazed), who was explaining the record-pressing process in the large "living room," on the factory's second floor, where the company frequently records live, limited-edition releases by local artists. "When a record is playing, grooves in the record are replicating those vibrations, and the needle is picking them up and amplifying those vibrations."

If that sounds simple, it both is and isn't, and serves as a good initial lesson in what is required to make an analog product a physical reality. Let's say we're talking about Taylor Swift's album *1989*. To transfer the vibrations of such songs as "Shake It Off" to a permanent, physical record requires several steps. First, Swift and her band record the album in the studio, where the edited tracks are mixed for balance by a producer, and mastered for the ideal volume by an audio engineer. The master recording is then played through a cutting lathe, which is basically a reverse record player that has a diamond-tipped cutting head instead of a needle. This cuts grooves into an aluminum disc that is covered in a semisoft black lacquer similar to nail polish. These grooves perfectly match the peaks and valleys of each song's sound wave, and are the small lines you see on a record.

Next, the lacquer master disc is transformed into metal stamper plates through a complicated process that involves chemical baths, bags of nickel nuggets, electric currents, and several stages of repetition. The metal stamper plates, for both the A and B sides of a record, are then affixed to a pressing machine. Essentially each machine squeezes a hockey puck–size "biscuit" of melted PVC with roughly 6,000 pounds of hydraulic pressure, imprinting the grooves of the sound waves representing Swift's songs into the vinyl, like a giant waffle iron. Each record takes approximately thirty seconds to press.

While the process sounds automated, it is highly variable and requires a heavy human touch. Everything from humidity to the particular mixture of metals in the stamper or the properties of a single batch of PVC can impact the quality of a particular record. URP is constantly inspecting the records that come off the presses for ticks, pops, or other "surface noise" that a needle would pick up, with microscopes, at listening stations, and with the human eye, and rejects up to 20 percent of the records it produces. These rejects are "dinked" in a machine that punches out the label, and then crushes up the vinyl to be melted down and pressed into new records.

"You simply can't standardize the process," Millar said, dinking a Metallica album that had a poorly adhered label, with a loud thud. "It's inexact. Every day we find a new problem. If you were a baker, this would be like changing your oven and pan every single day." Music is the biggest variable. Records have a finite amount of physical space for information, and the more you cram in there (say, a particularly loud heavy metal album, or bass heavy dance music), the more information has to be squeezed into those small grooves. This requires subtle tweaks at every stage of production.

"Every job I've ever had was in music," said Millar, who is slender and wry, just shy of forty, and speaks in an accent that's somewhere between his native Detroit, New York (where he lived for many years), and Nashville. He got his start at a record store, and eventually worked in marketing for Polygram, BMG, and Universal. He and his wife moved to Nashville in 2006, after Millar came for a Tom Waits show here and fell in love with the city. He was hired by United Record Pressing shortly after, and quickly became a key voice in the revenge of vinyl.

"I'm very representative of why the market got back into vinyl," Millar said. "I'd lived through vinyl, cassette tapes, CDs, and MP3s. I got all my music for free, and lived in a small New York apartment lined with CD cases." But when Millar got his first iPod, something changed. The music from his CDs could now live on various computers, so their physical presence didn't matter as much, but over time Millar missed the library aspect of his music: the art, the tangible feel and sight of them, the noticeable differences of sound quality among various albums.

"A lightbulb went on, and I realized that I had all of that with vinyl." Millar sold his CDs, and used the funds to buy back the vinyl versions of those albums. "Digitization is the peak of convenience, but vinyl is the peak of the experience," he said. Millar is quick to point out that he is no analog purist. He listens to digital music all the time: in the car, while jogging, or when his records aren't available. His wife even works for Warner Music as a digital production manager. "Digital is about making sure everyone has their music, and vinyl is the deluxe version, for the real music lover."

None of this explains the revenge of vinyl as an economic and cultural phenomenon. It is not as though Millar's "real music lovers" were a tiny tribe who suddenly grew more than ten times in size, as the vinyl record business has done in the United States since 2007, with similar numbers worldwide. What happened to the market for vinyl records before this sudden boom, and why was it growing so quickly now?

First, a bit of history: Commercial vinyl records were introduced in 1931 by RCA Victor, thanks to advances in polymer technology that made lighter, stronger, more durable records than the brittle wax and shellac gramophone discs of the time, which played at 78 revolutions per minute (rpm). They really didn't take off until after World War II, when Columbia unveiled the 12-inch LP record in 1948, which could play forty-five minutes of music at $33\frac{1}{3}$ revolutions per minute (rpm). A year later RCA came out with the 7-inch EP single, which played eight minutes of music at 45 rpm. These two formats, 12-inch albums and 7-inch singles, became the dominant vehicle for producing, buying, and playing the new postwar popular music at home, in jukeboxes, and on the radio.

Vinyl records had numerous drawbacks, including their size and weight, and the relative fragility of the vinyl surface, which tended to collect scratches over time that made records skip. They accumulated dust and static, took up a lot of space in stores and homes, and could warp in the sun. You couldn't play them in a car, let alone go jogging with them (not that anyone was jogging). Then, in 1979, Sony unveiled its first mobile cassette tape player, the Walkman, and four years later, the compact disc (CD). I vividly recall my father demonstrating our new, magical CD player in 1985. Its robotic tray slid open with an elegant whoosh, and he delicately placed a little silver disc inside. The house filled with crystal-clear sound (George Benson's *Beyond the Blue Horizon*, still one of my favorite jazz albums), and you could skip tracks at the push of a button. Here was a musical format for the emerging PC age, a clean, mysterious black box that somehow used lasers and digital processing to bring music alive. We had arrived in the future!

Vinyl record sales began slowly declining in the 1970s, as 8-track and cassette tapes ate at its market share. Singles peaked in 1973, with 228 million units sold that year in the USA; albums hit their zenith in 1978, with 341 million sold. The CD's rapid rise pushed the record off a cliff, with vinyl record sales falling by half between 1984 and 1988, and a continued decline thereafter. Album-length LPs suffered dramatically (they hit bottom in 1993, with just 300,000 albums sold in the United States that year), while singles, which were still used for jukeboxes and by DJs and radio stations, held out a bit longer. Still, the decline of the vinyl record continued well into the twenty-first century, as CDs gave way to MP3 downloads and the iPod. The year 2006 was the vinyl record's nadir. Worldwide album sales of new records totaled just 3 million units that year, and in the United States just 900,000 records sold, roughly a quarter of what Disney's *High School Musical* soundtrack did in combined CD and download sales that year alone.

Mark Michaels, the owner and CEO of United Record Pressing, had purchased the business in 2007, after a successful career in global management consulting and private equity. An amateur music collector, Michaels felt URP would make for a nice, steady business that could generate cash flow over time. "The commercial vinyl had declined to a tiny, tiny fraction of the industry," he recalled over the phone from his office in Chicago. But record labels still pressed promotional records for each new single, and this was the core of URP's business. "That part was small and stable," Michaels recalled, but he was unaware how suddenly the promotional vinyl business would fall off a cliff. "I didn't see that coming, but the labels were looking at business models and realized that giving away twenty thousand records may not be the best thing for the bottom line." When the recession hit a year later, it nearly killed the company. Michaels begged lenders to be patient, and laid off most URP employees. Many of its machines sat idle.

Vinyl records were, by any objective metric, dead. As one veteran record executive told me, at that point they were a statistical anomaly, a rounding error on the balance sheet of record companies, a fraction of a fraction of a percentage point of sales. By 2007, the music industry was deep into its struggles with digital downloads and piracy, and the future, though turbulent and uncertain from a revenue standpoint, was clear: digital, disembodied music, delivered wirelessly anywhere, anytime. CD sales were in continued free fall and even paid digital downloads were beginning to dip as streaming services, such as Spotify, grew in popularity. Physical music was on its way out. Vinyl was just the first victim.

Then came its revenge.

"By all accounts, ostensibly 'replaced' by the industry with a 'better' product, vinyl should have been dead by now, or at best confined to museums and antique stores as quaint incunabulum," wrote Dominik Bartmanski and Ian Woodward in their fascinating 2015 book *Vinyl: The Analogue Record in the Digital Age*. "But something else happened instead. . . . vinyl saw a socially broader renaissance exactly at the time when the digital revolution seemed complete."

According to the Recording Industry Association of America, LP album shipments in the United States alone grew from 990,000 in 2007 to more than 12 million in 2015, with annual growth rates of more than 20 percent. Various sources reported that by 2015, vinyl sales generated nearly a quarter of revenue derived from music sales, surpassing ad-supported streaming, as paid downloads and CDs continued to decline. New vinyl record sales generated $346.8 million for the music industry in 2014 alone, and likely many times that for sales of secondhand records, which still make up the bulk of vinyl sales.

Since its low point a decade back, the vinyl record has grown rapidly, dramatically, and steadily. It is a stunning reversal. For some reason, people have purchased more vinyl records (new and used) in the past ten years than they did in the prior twenty combined.

Why?

First, the vinyl record never died. New album sales plunged to their low from a peak where vinyl records made up nearly 100 percent of music sales, but the existing records in the market, billions and billions of them, were a physical reality that simply did not disappear. They hibernated in shelves, crates, and boxes in record stores, flea markets, and basements. Every turntable that existed up to that point was still a physical reality. The infrastructure was dormant, but largely functional. "There was always a good market," said Heinz Lichtenegger, CEO of the Austrian turntable manufacturer Pro-Ject Audio Systems, who began his own business in 1991, selling mid to high-end turntables at a time when other turntable manufacturers, such as Technics, stopped producing them. "From that first day I always had back orders," he recalled. Pro-Ject's core consumers were made up of fidelity-obsessed audiophiles and anticorporate punks, German jungle DJs, and moneyed collectors. They tended to pay more for records, forming a protective niche that allowed many record shops, pressing plants, and turntable companies to stay in business during the worst years.

There were some pockets of growth for vinyl, especially with underground genres, including punk, hip-hop, and dance music. Ton Vermeulen purchased a Sony plant outside Amsterdam in 1998, which pressed dance music records for the huge European nightclub market. "When I entered the building, we were growing very rapidly," he said, "not because vinyl was growing, but because a lot of pressing plants were going out of business." Vermeulen estimates that even back in 2000 there were tens of millions of records being pressed globally, largely for the club market, which needed new singles, beats, and tracks for DJs to spin with.

Second, digital helped save the very analog record it nearly killed. As record stores closed, the market for vinyl grew increasingly niche and vinyl fans turned to the Internet to buy and sell records. Millions of albums were auctioned on eBay, sold on Amazon, and traded on the massive online marketplace Discogs. Meanwhile, digital music's advantages became its disadvantages. The advent of MP3s was harder on CDs than on records, and CDs (which offered no real sonic or aesthetic advantages over digital files) became an obsolete way station to the more mobile, space-saving MP3s. Because it could be copied infinitely, without a loss of quality, an illegally downloaded album was in no way different than a legally purchased one. Napster laid that truth bare in 1999, and the music industry never recovered. Once music was divorced from any physical object, its supply so vastly exceeded demand that people simply refused to pay for it. Suddenly, an album was no longer a desirable object worthy of consumption. All digital music listeners are equal. Acquisition is painless. Taste is irrelevant. It is pointless to boast about your iTunes collection, or the quality of your playlists on a streaming service. Music became data, one more set of 1's and 0's lurking in your hard drive, invisible to see and impossible to touch. Nothing is less cool than data.

Meanwhile, the previous disadvantages of vinyl records now became attractive. Records are large and heavy; require money, effort, and taste to create and buy and play; and cry out to be thumbed over and examined. Because consumers spend money to acquire them, they gain a genuine sense of ownership over the music, which translates into pride.

Vinyl records reacquired a counterculture cachet, which propelled them back into the core of youth culture. "Kids started buying records," said Tom "Grover" Biery, a music executive in Los Angeles who was an instrumental figure at Warner Music when vinyl began to grow again. "As iPods and Facebook became their parents' stuff, kids began searching for something different, because it wasn't cool once your parents did it . . . just like rock and roll. And vinyl was not their parents' stuff anymore."

A 2015 research report in the United Kingdom found that the main consumers of vinyl records that year were 18- to 24-year-olds, and research group MusicWatch noted that more than half of vinyl buyers were under 25. Not ageing, retro hipsters. Not crusty old dudes. Kids who are discovering records for the first time. While baby boomer parents gushed about their new iPads and Spotify accounts, their children were dusting off old turntables and buying new albums for cold hard cash. Vinyl records went from a retro fetish to a cool new consumer good. Turntables appeared in ad campaigns, fashion magazines, and boutique hotels. "There wasn't a couple days going by [since I opened in 2011] where I wasn't showing kids in their early twenties how to put the needle on the record," said Craig Brown, who owns the record store Heights Vinyl, which opened at the end of 2011 in Houston, Texas. "These are first-timers. That's the market."

The third major reason for the revenge of vinyl was more deliberate: Record Store Day. This annual celebration of vinyl record retailing occurs the third Saturday of each April, and appears to be the final push that kicked the vinyl record revival into the mainstream. In 2007, a small group of independent music store owners who belonged to a coalition called the Department of Record Stores were having their annual meeting in the basement of the Sound Garden, a store in Baltimore. The conversation turned to the state of the business, and all the owners told the same story: each had survived brutal, cutthroat price-gouging competition throughout the 1990s from such big-box retailers as HMV, Tower Records, and Virgin, and each had weathered the downturn in CD sales over the previous decade. But despite all this, all their stores were doing well and making money.

"We were opening multiple stores, posting twenty percent annual growth," said Chris Brown, the CFO of Bullmoose, a chain of record stores in New Hampshire and Maine with eleven locations. Around that time, Bullmoose was doubling the size of its existing stores, knocking down walls to make room for more vinyl records and books. "It was totally counter to what everyone was reporting on," Brown said. Even though these stores were profiting, the public perception was that they were on death's doorstep. Customers regularly came in and asked, "How are you doing?" with great sympathy. Record stores had become largely irrelevant to all but the most ardent music fans, and that affected their very identity.

"Employees used to fight over new releases," recalled Michael Kurtz, who ran the Department of Record Stores. Early access to new music was one of the main reasons someone worked in a record shop. "Now that was all available online before in-store. Our own employees [did] not give a shit. The amount of younger women who wanted to work in our stores had been reduced to zero," Kurtz said. "We were becoming the comic book store nerd."

Eric Levin's ears perked up when someone made the very same comic book nerd joke during the meeting at Sound Garden. Levin's business (Criminal Records, in Atlanta) had recently hosted an event for Free Comic Book Day, a promotion by the comics industry that proved incredibly popular. Why not do a similar day for record stores, to prove to the public and media that the record store was alive? "I'd read that we were dead; worse than buggy merchants," Levin said, characterizing popular perception of record stores at the time. "But in my store, we were really having fun, employing people, insuring people, and making money. Why was the press so negative? The reporters wanted proof," Levin said. "They wouldn't believe I was successful. It didn't fit in with their norm. It didn't prove what they were reporting. It was unusual: 'How could you be doing good if music is free? If Tower is closing? If Best Buy is hurting?'" Even though most of the stores were still selling far more CDs, Levin insisted that vinyl had to be the focus, because it was the way to get people to talk about the stores themselves.

In the months leading up to the first Record Store Day, Kurtz paid a visit to Tom "Grover" Biery in Los Angeles, who was then general manager of Warner Music. Biery traced Warner's renewed interest in vinyl to an encounter with Neil Young, who came into the office in the early 2000s to listen to a digitally remastered greatest hits compilation Warner was set to release on CD. Young was so disappointed in the sound quality that he gave an impassioned speech about how no one was standing up for the music and artists in the end product anymore. "That's when we took a conscious decision to really focus on vinyl," Biery said. "It was never about revenue. It didn't matter if we were gonna make or lose money, it was a good thing to do for our brand awareness." Warner, like all record companies then, was hemorrhaging money and didn't have much to lose in dusting off an old but proven format.

Slowly, Warner began releasing small pressings of select albums, aimed at the audiophile and collector market, or as merchandise that could be sold on tour. These albums were pressed on heavier vinyl, to appear more substantial (and, some say, sound better), and included music by Wilco, the White Stripes, and others. After Kurtz told Biery about the idea for Record Store Day, Warner agreed to give some modest financial support for marketing. More important, Warner would provide the record stores with limited-edition releases by artists, including Death Cab for Cutie, R.E.M., Vampire Weekend, and Jason Mraz, which would only be available on Record Store Day at participating stores. Some three hundred stores around North America and the United Kingdom participated in the first Record Store Day, on April 19, 2008. The coup was Warner getting Metallica to serve as official Record Store Day ambassadors, who signed autographs at Rasputin Music, in Mountain View, California.

It was a triumph. "It's the first time we've seen lines of customers waiting outside our stores before we opened, ever," Kurtz recalled. "That never happened before." Many stores reported sales up to 50 percent above average. The press came out in droves, and the positive media coverage was exactly what Levin had hoped for. Record Store Day kept expanding. By 2009, the single day's sales at many stores dwarfed their numbers on Black Friday and Christmas. More stores, more concerts, and more exclusive releases followed. Collectors waited overnight in lines to snap up limited Record Store Day albums, which they sold online, almost immediately, for many times their face value. That brings no shortage of complaints from store owners and fans, but it is a symptom of success rather than failure. Record Store Day is now a global celebration, with record stores, artists, and labels participating worldwide. Last year, one of the hundreds of record stores to participate was Dund Gol, which had recently opened up in Ulaanbaatar, Mongolia.

Just walk around any major city, or even larger towns, and record stores are opening and expanding at an incredible pace. In the United Kingdom alone, the number of record stores reached a five-year high by 2015's Record Store Day, with forty new shops opening in the first four months of that year around Britain, 50 percent more than opened during all of 2014. In vinyl-heavy Berlin (easily the world's biggest market), there are now more than one hundred record shops,

and popular ones, such as OYE, are expanding with multiple locations. Even large chains are getting back into vinyl. British retailer
HMV, which dominated the CD era, emerged profitably from bankruptcy in 2015, driven largely by vinyl sales. Here in Toronto, a new
record shop opens every two months.

"We tapped into something by accident," said Kurtz, who now
devotes most of his time to managing Record Store Day. "It was a
need for people to connect with the community, like a music festival.
A chance for people to gather and celebrate music. We created an excuse for a party, but were surprised when people showed up."

At United Record Pressing, all of this trickled down through the
business. Although the promotional singles market never recovered,
the growth of orders for album-length LPs began stirring in late 2009,
and moved steadily upward from there. Michaels had smartly repositioned the company for this surge, after reading a *Billboard* magazine article about the nascent growth in vinyl records for the digital
generation, and had reengineered United Record Pressing to focus on
higher-quality album production, rather than churning out inexpensive singles.

"By 2012 all of a sudden you saw some tipping point had been
crossed and you saw single digit growth become double digit growth
at an accelerating pace," said Michaels; "2013 was exceptional, and
2014 the market grew beyond our ability to meet the demand." URP
hired more workers, bought more presses, and refurbished old ones.
Even with this, the factory still couldn't meet demand running non-
stop, six days a week, especially in the months leading up to Record
Store Day. Michaels had to build a second URP factory in 2015.

The forty-odd record-pressing plants around the world are all
now running at full capacity. "The first three months of 2014, I pressed
the same number of records I did in 2013," said Ton Vermeulen of Hol-
land's Record Industry plant. "In April, the shit hit the fan. In September I started a second shift." After decades of vinyl plant closings,
new record-pressing operations have opened in Memphis, Michigan,
Louisiana, Alberta, London, São Paulo, and a dozen more locations.
Gotta Groove, which opened in Columbus, Ohio, in 2009, was initially pressing around fifteen thousand records a month. By 2014

it was up to seventy thousand. "If the capacity really is fixed," said Gotta Groove's co-owner Matt Earley, "and the demand is continuing to grow, and will, it just seemed to be the right business decision."

Auctions for the limited number of record presses in existence, which can fetch up to $80,000 apiece, are fierce, and people are hunting these antiquated machines down as far away as Zimbabwe and Trinidad. When Chad Kassem opened his Quality Record Pressings plant in Kansas in 2011, he purchased his presses from an old EMI plant in Los Angeles and another in London. When I asked him how he finds presses, he said: "Any way you fucking can, is how you do it!" Earlier last year, Kassem found thirteen presses that had been forgotten in a warehouse in Chicago, the record man's equivalent of striking oil. But like an untapped well, each newly acquired press needs tens of thousands of dollars of precisely machined parts, maintenance, boilers, coolers, stampers, molds, and other work to get to the point where it can press a record. Last year, a Germany company called NewBilt Machinery and a Toronto company called Viryl Technologies finally began manufacturing the first new vinyl presses to hit the industry in half a decade. Orders went to United Record Pressing.

Turntables are another key barometer of vinyl's revenge. Although Heinz Lichtenegger continued to steadily grow Pro-Ject Audio Systems throughout the 1990s, in the 2000s he sensed various national markets waking up to vinyl: first the United Kingdom, then Germany, Italy, Scandinavia, and finally, around 2010, North America. "It totally changed," he said. "Magazines and newspapers were writing about vinyl and turntables and our turntable business doubled from 2011 to 2014." He currently has a back order of thirty-five thousand turntables in production, and estimates the global market for turntables is around 5 million units a year. That is enough of an opportunity to bring companies who previously produced turntables, such as Pioneer and Sony, back into the business, while such mainstream retailers as Target and Walmart have stocked their stores with record players.

Most often, those turntables are inexpensive models made by Crosley, a Louisville, Kentucky–based brand that used to make mini-jukeboxes and other retro-themed home electronics. Crosley has been manufacturing a turntable since the 1980s, often paired with CD players and radios in vintage-looking cases. "Our business was driven by their kid seeing the item in the Sky Mall catalog and saying, 'Gosh, for ninety-nine dollars this would be a great Christmas present for Dad,'" said Crosley's VP of sales and marketing at the time, Elizabeth Braun (who has since moved on from the company). But as the older consumer became more savvy with digital CD burners and iPods, their nostalgic interest in records waned, right as their grandchildren began to get into vinyl. So, Crosley shifted its focus to that consumer.

Crosley now makes more than two dozen different styles of turntables, from portable units with built-in speakers to branded editions featuring the image of the boy band One Direction, or The Ramones. They sell over 1 million turntables each year. "Our new demographic has never seen a turntable, never touched a record," Braun said. The previous day she had been on a conference call with a younger Urban Outfitters marketing team member (the chain now sells more vinyl and turntables than anyone else in America), who asked Braun what the little lines on the records meant. "I had to tell her those are the songs," she said.

From the economic point of view of the music industry, the vinyl record revival is something that is easily dismissed as a quirky footnote. Although vinyl album sales have grown dramatically since 2007, they still represent less than 10 percent of all music sales, and pale even compared to current CD sales, which still sell more than three times as much. But look deeper into vinyl's market, and there's more at play. "For a long time, I was in that camp where every time I saw an article on vinyl buying, I'd say, 'Yeah, so we went from five people to seven people . . . so what?'" said Russ Crupnick, founder of the market research company MusicWatch. But the key to vinyl is its high price. When CD sales began their long decline in the late 1990s, with the advent of digital downloads, the labels responded by slashing pricing to the point where CDs were barely making a profit. New vinyl buyers, in contrast, aren't as price sensitive. They are happy to pay $20 or more for a copy of Taylor Swift's *1989*, or twice that for a Record Store Day special release, because they get something substantial in return: an asset they can hold in their hands. "You could sell a lot less vinyl to a lot fewer people and make higher profits," said Crupnick. Compare that to digital downloads, where he estimates a label needs to sell over 127,000 singles to break even on the production of an album. "The average price of that CD is six dollars wholesale, and you maybe get sixty cents [wholesale] on iTunes," Crupnick said. "But suddenly on vinyl, you can bump up to ten or twelve dollars wholesale . . . that's significantly more profitable than other models."

Billy Fields, VP of sales and account management at Warner Music Group, is skeptical that new vinyl records will ever top $1 billion in sales, but he sees no slowdown in a format that is becoming a significant driver of profit. "Revenue per unit on LPs is far higher than anything we can sell. While there's profit baked into everything we do, I say that it's probably a really solid, high double-digits profit margin. Not as good as we receive on a digital download, but the revenue per unit is much higher." And while the total revenues of digital downloads and CD sales are greater than for vinyl records, those two formats are in steady decline, due to the rise of streaming services, while vinyl continually expands. "Vinyl will grow because young people are making an investment," said Jeff Bowers, an executive with Universal Records. "They're buying a turntable, and a turntable only plays records."

Then there are musicians. Despite the wealth of great music created since the MP3 and iPod came out, and the terrific opportunities for artists to speak directly to fans and distribute music online, the post-CD, second generation of digital music has been a horrible time for musicians hoping to make money off recorded music. Each new technology to acquire music that has come along since 2000 has introduced new ways to screw musicians out of the proceeds for their art. Streaming services may talk about compensating musicians fairly, but whenever anyone discloses real figures, the reality is grim. In articles, blog posts, and royalty statements posted online, musicians, ranging from well-known songwriters to fairly popular up-and-comers, talk about incredulous figures. Songwriter Aloe Blacc told Wired that his reward for cowriting one of the most streamed songs in recent years (the foot-stomping dance hit "Wake Me Up!" by Avicii) from the music service Pandora (which made nearly $1 billion in revenue in 2014) was less than $4,000, far more than the $16.89 Cracker's David Lowery got from Pandora for a million plays of the song "Low." In 2015, both the US and UK record industries reported that more money was generated from vinyl sales than from advertisingsupported streaming services, such as YouTube, the free versions of Spotify and others. "With streaming they get paid in pennies," Michael Kurtz said. "With vinyl they get paid in dollars. It's a completely different world. If an artist and label knows what they're doing, they're going to make hard cash."

With chump change like this, it's no surprise that major artists from Radiohead's Thom Yorke to Taylor Swift have refused to work with certain streaming services. And the reason is clear: the streaming services have proven technology, but unproven business models. Most are funded entirely by venture capital investors or by parent companies, such as Google and Apple, and on average they spend more money than they earn, giving away the bulk of music for free, in the hope that a significant percentage of listeners will purchase a premium subscription, such as the $10 a month I paid for access to Rdio's full catalog. To become profitable, these streaming services need massive numbers of paid users to opt into subscriptions. With so much music available for free online, and new services popping up daily, hitting those numbers is a tall task. As I was editing the second draft of this chapter, I heard that Rdio had gone bankrupt, failing to turn a profit even after raising more than $175 million (plus $150 from me over the course of the year I subscribed). I phoned my friend Adam Caplan, a fellow Rdio listener, and broke the news. "Oh, man," he said with frustration, noting that he'd just synced up his whole music collection with the service, and now had nothing to show for it. I told him the money would be better spent on more records. "Yeah," he agreed, "the future of music is the past."

None of this even touches on the market for secondhand vinyl, which isn't tracked by the industry, and still makes up the bulk of records traded, sold, and played today. In 2015, the online music marketplace Discogs sold over 5 million records alone, representing just a slice of the secondhand market. For record stores, the profit margins on a used record are much higher than on a new one, and the supply is only growing, as more records are pressed, sold, and resold. With each new vinyl listener who plugs into a turntable, the value of used records increases. Darren Blase, who owns the store Shake It Records in Cincinnati, Ohio, says that while new records carry a 40 percent profit margin, used records can sell for double their cost, or more, and the price of used records is increasing each year. "We're able to sell a lot of stuff that we couldn't before, like that first Boston album," Blase said, noting a frequent bargain bin selection, lending hope that all those unsold Herb Alpert and the Tijuana Brass records out there will soon find a home.

———————•———————

*D*uring *my tour of the* United Record Pressing plant, one musician's albums took up a disproportionate number of the records being pressed at any given time. This was Jack White, the Detroit-born, Nashville-based, guitar-shredding rock god behind the White Stripes, the Dead Weather, the Raconteurs, and his own solo work. Beyond his prowess on the stage, White is the most vocal musician advocating for vinyl and analog music today, and a visionary businessman when it comes to making and selling records.

"With vinyl, you're on your knees," White told *Billboard*. "You're at the mercy of the needle. You watch the record spin and it's like you're sitting around a campfire. It's hypnotic." White has said that there is no romance in a mouse click; that physical, analog recording technology preserves music and sound far better over the course of history than quickly obsolete hard drives. He has recorded one of the top-selling records of the entire vinyl boom, *Lazaretto*, an album so packed with wild quirks and design features—a hidden hologram, a secret track on the label, an A side that plays from the inside out (which I didn't figure out for a full six months)—it serves as analog's battle cry.

If White is the industry's Willy Wonka, then Third Man Records is his chocolate factory. In 2008, White converted a former auto repair shop into a psychedelic, analog wonderland that looks as if it was designed by Tim Burton. Each surface is painted either glossy black or in shocks of primary yellow, red, and blue. Everyone working there wears the company's signature black and canary yellow colors. Third Man has a vintage pharmacy soda fountain, a stuffed giraffe's head, old quirky cars, and motel signs. There is a concert venue where artists have performed live to the cutting lathe, or recorded with White as producer, ranging from local independent bands and Aziz Ansari to such legends as Beck, Loretta Lynn, Wanda Jackson, Jerry Lee Lewis, and Willie Nelson. All these records fall under the Third Man label, including all of White's music, live performances recorded at Third Man, rare reissues (including Elvis's first recording and a Carl Sagan single that has sold over 10,000 copies), and emerging artists, such as Kelly Stoltz, Seasick Steve, and the Haden Triplets. There is a quarterly mail-order record club, called the Vault, and a small retail store that sells records, novelties, and Third Man–branded merchandise, including black and yellow turntables by Pro-Ject and Crosley. There's even a vintage 1930s record booth, where you can cut your own 7-inch single where Jack White recorded an entire album of Neil Young singing covers.

Third Man's motto is unapologetically analog: "Your Turntable's Not Dead."

"Vinyl is an identity," said Ben Blackwell, a musician (and also White's nephew) who runs Third Man. "We have done over two hundred and fifty releases in five years and pressed over one million pieces of vinyl," Blackwell told me in the fall of 2014. Although Third Man also puts out its music as digital downloads and on streaming services, the focus is on vinyl. For Third Man, analog is not simply the pinnacle of the artistic experience, it's good business. It not only sets the company's products apart from a merchandising and profitability sense, but it shapes the entire visual and sonic aesthetic of the label, which is a direct translation of White's unabashed love of analog music.

"Analogue is the medium of all the kinds of music that I am really fond of," White told the recording industry magazine *Sound on Sound*. "When you are recording and producing, you are aiming for something and if you want vibe, warmth, soulfulness, things like that, you will always be drawn back to analogue. . . . the actual sound of analogue is ten times better than that of digital."

Up to now I have avoided discussing sound's role in the revenge of vinyl. There's a reason for this: as soon as the conversation turns to a comparison of the different sonic qualities of music formats, it becomes loaded with technical arguments on compression rates, speaker frequencies, and dynamic ranges. Audiophiles can spend their lives chasing the perfect weight to balance their turntable's tone arm, and the web is filled with forums discussing whether anyone can detect the difference between a WAV file and an MP3 on five competing headphone brands. Digital music takes an analog sound wave and translates it into 1's and 0's, inevitably sacrificing chunks of information, and sound, in the process. Usually, digital files are compressed to a smaller size to make them easier to download and stream, and their volume levels are jacked up to compensate. But none of that really matters to the vast majority of music listeners, who aren't really that concerned about sound quality.

Where sound does matter, and where analog music is experiencing a parallel revenge to the resurgence of vinyl, is in the studio. Until the 1990s, pretty much all music was recorded to magnetic tape. These tapes were linear and limited in their size, so if you recorded something, the only way to change it was to record over the top of it, or to literally take a razor blade and cut the tape. Digital synthesizers and other musical equipment have been around since the late 1960s, but in 1991 the first Pro Tools audio editing software was released. For the first time, music producers and engineers were unconstrained by the limits of tape. They could cut and paste certain solos or sounds in the computer, and drag them where they pleased with a mouse. Musicians could listen to a track, decide what parts they liked, try something new, and if it didn't work out, just click Undo. There were no more mistakes.

By the early 2000s, digitally recorded music had become the industry standard. As hard drive capacity and processing power dramatically increased, Pro Tools became so inexpensive that musicians were able to set up studios in their own homes. Other companies created complementary programs called plug-ins: Auto Tune corrected out-of-tune vocals by locking them to the nearest note in a key, the Ultramaximizer automatically increased the music's volume, while others replicated studio effects, such as reverb and echo, that were previously achieved with bulky equipment and complicated processes. If you wanted an echo before Pro Tools, someone had to perform at one end of a long concrete echo chamber, with a microphone placed at the other end.

Just as the digital dominance of the recording studio seemed complete, analog had its revenge. Musicians, producers, and engineers searching for the sound of the music that inspired them—roots Americana, blues, and classic rock—began thinking about how the process of recording affected the sound. These artists, including White, Dave Grohl, and Gillian Welch, began experimenting with old tape machines and vintage studio equipment, returning to the analog methods they'd once used. Critics and fans noted that these albums sounded different—more heartfelt, raw, and organic—and the industry began to take notice. In 2011, the Foo Fighters won a Grammy for *Wasting Light*, which was largely recorded in Grohl's garage with analog equipment, in as few as three takes. "You have the ability to completely manipulate and change the performance with the digital stuff. You don't really have that with analog," Grohl told an interviewer. "I don't want to know I can tune my voice, because I want to sound like me," he said.

Nashville has since become a magnet for musicians seeking this analog sound, and there are studios around town that specialize in capturing it, including those owned by Jack White and the Black Keys. The largest is a windowless former vinyl record–pressing plant out in the city's west end, packed with every vintage instrument and recording gizmo you can imagine, called Welcome to 1979. "You can't fight the analog vibe here," said the studio's co-owner Chris Mara, as we sat down behind a huge wood-paneled recording console in a room packed with guitars, racks of vinyl records, tapes, and oriental rugs. "It's so seventies!"

Mara had worked as a freelance recording engineer for over twenty years, initially training with veterans of Alabama's legendary

Muscle Shoals studio. When he went out on his own he faced a choice: spend over $50,000 to buy the latest Pro Tools setup and create a home studio, or use the analog equipment he already owned and build a real studio. In 2008, Mara borrowed against the value of his pickup truck, and opened Welcome to 1979. His business has doubled every year since. "I think the sound quality is one of the smaller reasons why people use analog," he told me. "Really, it's the process. This stuff," he said, sweeping his arm across a laboratory of painfully restored, forty-year-old gear, "was designed by musicians primarily to record them. Pro Tools was designed for engineers."

Musicians and bands come to Welcome to 1979 and other analog studios because the endless options, variations, tweaks, and plug-ins of digital studios create a moving target of unachievable perfection. You never have to make a firm decision with digital, because you can always drag the mouse to change the sound just a little bit more, and just click Undo if that doesn't work out. Mara saw artists frequently burning out in digital studios after too many takes and edits. By contrast, analog presents a vastly more limited workflow: you play music, it gets recorded to tape, the tape is played back, and you decide whether it's good or you want to record another take. This may sound simplistic, but remember that most of the greatest music you know, from Miles Davis's *Kind of Blue* to the Beastie Boys' *License to Ill* was recorded this way, sometimes in a single session. "People think limitations are a bad thing," Mara said. "But it moves the process forward, in a good way. You can easily get lost in the process. It's easier to stick to the plan when you have limitations."

After meeting Mara, I drove down to Franklin, Tennessee, to talk with Ken Scott, one of the most legendary recording engineers in popular music. The first record he ever worked on, at just sixteen years old, was *Hard Day's Night*. Scott stayed behind the console at EMI's Abbey Road studio for the bulk of the Beatles' recording career, and has worked with David Bowie, Pink Floyd, the Rolling Stones, Lou Reed, Elton John, Devo, and Duran Duran, to name a few. He learned to record with "all hands on board," when the fingers of producers, engineers, musicians, and even the kid fetching coffee were literally commandeered to create a desired sound live in the studio by pushing switches and dials on the console. This chaotic process often led to "happy accidents," which became integral to the final recording. During the mixing of "I Am the Walrus," Scott was pushing faders, John Lennon banged on the piano, Paul McCartney tweaked a tape reverb machine, George Harrison made sounds into a microphone, and Ringo Starr was flipping through the radio, adding to the chaos. Suddenly, Ringo landed on a BBC broadcast of *King Lear*, and the contrast of Shakespeare with this audible acid trip made music history. That kind of spontaneous, arcane improvisation could only happen in an analog studio.

Over the previous decade, Scott had witnessed the unmistakable impact that digital recording technology had on the studio sessions he worked on. He had sat there as producers made rock-and-roll legends record fifty-nine guitar solos in a row, so they could edit bits of each together, instead of asking the musician to give his best take. He observed the greatest singers in the history of popular music having their vocals distorted through Auto Tune, and shook his head when bands opted for a drum machine instead of the Grammy-winning drummer hired for the session.

Scott wasn't against digital equipment. Some of the earliest digital consoles had been used to record parts of *Abbey Road* and *Dark Side of the Moon*, and certain musicians, such as Daft Punk and Kanye West, based their sound on digital production. A computer was just a tool, but Scott felt the computer was being overused, and that musicians, producers, and labels were crafting their music to suit the technology's biases. They recorded safe, vanilla sounds tailored to edit in Pro Tools. They got lazy. "A lot of recording has lost its soul," Scott told me, as some bubblegum country pop played in the Starbucks where we met. "It comes from the head more than the heart. Done by looking at a screen, instead of *listening*." No one took risks anymore. Take "Five Years," the opening track on David Bowie's masterpiece *The Rise and Fall of Ziggy Stardust*, an album Scott had engineered. "The end of that song is so emotional," Scott said. "Bowie literally had tears streaming down his face as he was singing it." Today, an audio engineer would smooth out those cracks in Bowie's voice with

Pro Tools as he sang his heart out, even though that shaking voice is precisely what makes the song's finale so arresting.

Scott, like several others I met, feels that the renaissance of analog recording goes hand in hand with the vinyl revival. The same musicians and bands who want their music on vinyl crave the sound of vinyl's heyday, and appreciate the process that goes into capturing it. This includes such bands as the soulful southern outfit Alabama Shakes, whose debut was recorded at the analog Nashville studio the Bomb Shelter, and big names, such as the Arcade Fire. Even Ryan Adams's album-length cover of Taylor Swift's *1989* was recorded entirely in analog. Initially confined to rock and roll, the interest in analog recording is now steadily moving into other genres. Two of my favorite albums from 2015 were both proudly analog: *Black Messiah*, a triumphant, ripping comeback by the soul singer D'Angelo, and *Sour Soul*, a collaboration between Wu-Tang Clan veteran Ghostface Killah and the instrumental funk band BadBadNotGood. Both are the kind of record that immediately stop you in your tracks.

On my last night in Nashville, I bellied up to the bar at the Stone Fox with a burger and beer, and watched as a local band called Promised Land Sound set up on the small stage. It was your typical Nashville rock outfit: four young guys in their early twenties, with varying lengths of hair (facial and otherwise), clad in denim and plaid. The bar was half-empty, and most people were focused on their phones.

Then Promised Land Sound started to play. The band immediately let loose a torrent of wah-wah–heavy, psychedelic country garage rock. There were tastes of the Band, the Byrds, and the Allman Brothers in there, but it was clearly its own thing, a pure Nashville sound. Halfway through that first song I took my eyes off the drummer and noticed heads bobbing, feet tapping, and smiles breaking out all around the bar. All the smartphones had disappeared into bags and pockets, chased away by the sheer energy of the band's music. It commanded your attention so fully, so mercilessly, that you either had to submit or flee.

This was what drew in teenagers who had grown up with iPods and never touched an LP to rush out and buy their first turntable. This was why vinyl records and analog music had their revenge.

Because, to paraphrase Mark Twain, a great live band is a bolt of lightning, and an iPod is a lightning bug.

After the set, I went up to the guitarist and asked if the band had brought any records to sell.

"Yeah, man," he said. "Of course."

2

The Revenge of Paper

If you ever attend Milan's Design Week—a sweeping furniture fair, art festival, and Prosecco-soaked party that takes over Italy's financial capital each April—you will need several essentials to fit in with the global trendsetters in attendance. First, your glasses. This is a design crowd, so the options are polarized into two camps: ultra-minimal frameless spheres that hover over your face, and inch-thick acetate beasts. Next, your clothes. Leave ankle-snapping heels and impossible dresses to the fools at Fashion Week; the name of the game here is the ideal mix of form with function, such as a limited-edition pair of Converse clad in sharkskin. Also scarves. You can never wear too big a scarf at Design Week, even if a sudden heat wave descends on the city. Finally, there's your kit. If you carry a bag, it needs to be small, streamlined, and slung over the shoulder. In one hand, firmly grasp the latest iPhone, which you'll use to take photos of new chairs and kitchen tile installations, stay in touch with colleagues, and navigate your way between parties and events all over Milan.

In the other hand, you will carry a black Moleskine notebook. Perhaps you will bring the same, lovingly worn Moleskine back to Milan year after year, or maybe you will unwrap a fresh one at the start of every Design Week. You will jot quick notes and sketches in it as you tour showrooms and exhibits, and at the end of each long day, you will open it on a café table, order a cold Peroni and a plate of mortadella, put pen to paper, and digest all the exhibits, products, and far-reaching designs you've seen since the morning. You might start by writing lists of words, maybe descriptive adjectives, or a few lines of insights that tease out this notion you have forming about the organic desire for analog experiences in a hyperdigitized age.

After a page or two, you'll stop, take a sip of your beer, and look around. You'll draw inspiration from the pale gold of the beer, the beautifully irregular spots of the mortadella, the bleached ceramic plate it sits on . . . and suddenly you'll be designing your own textile line, filling pages almost as fast as you can turn them, with sketches and words and feverishly added details.

Yes, I am being romantic, and perhaps mortadella isn't a traditional muse. But this is exactly what I saw at the end of each day during my visit to Milan in the middle of Design Week. Moleskines . . . hundreds of them, thick with the ideas being written and sketched in them by the most talented and driven design minds in the world. Old designers, young designers, European designers, and Asian and Latin American designers—the Moleskine notebook is their common tool. You certainly do not need to visit Milan to witness this. Just step into any coffee shop, pretty much anywhere in the world, and you will find Moleskine notebooks resting on tables next to lattes and laptops. With their rounded corners, elastic closure, and ivory paper, they are instantly recognizable. Which is why I see them everywhere: in the hands of teachers, on my doctor's desk, in the bags of computer software programmers. Nearly everyone I interviewed for this book pulled out a Moleskine notebook at some point, or had one sitting nearby. For a thoroughly analog object, the Moleskine is one of the iconic tools of our digitally focused century.

I am fascinated by Moleskine for several reasons. Paper was the first analog technology to be seriously challenged by digital. Even before the rise of the personal computer, the term *paperless office* had emerged as an obsession of the business world by the late 1970s. The promise of a computer reducing the cost, labor, and space associated with printing, storing, and organizing so much physical information was powerful. Office workers were called "paper pushers" for a reason, and less time pushing paper meant more time, and money, devoted to other work. Although the fully paperless office has yet to come to fruition, certain paper artifacts are now largely digital. E-mail, texts, and PDFs have taken the place of memos, telegrams, and faxes. When I entered university in 1998, I was one of the few students to take class notes on a miniature laptop. Today, screens overlook college lecture halls. Paper hasn't disappeared, but it is no longer dominant.

Paper is also the oldest analog technology to be seriously challenged by digital. Vinyl records had been around less than forty years when the CD came out, but paper has existed, in one form or another, for thousands of years. It is the backbone of the economic, cultural, scientific, and spiritual core we call civilization. Even the bound notebook, the core technology behind Moleskine's main product, predates European settlement of the Americas. Our relationship to paper is older, deeper, and more varied than any other analog technology out there. Understanding where paper's advantages lie, where it has struggled to compete with digital technology, and where it has crafted a new sense of identity is key to understanding how the Revenge of Analog is playing itself out.

The revenge of paper shows that analog technology can excel at specific tasks and uses on a very practical level, especially when compared to digital technology. While paper use may have shrunk in certain areas since the introduction of digital communications, in other uses and purposes, paper's emotional, functional, and economic value has increased. Paper may be used less, but where it is growing, paper is worth more.

No product better captures this niche than the Moleskine notebook and the company behind it. It is *the* defining paper object and brand of the Internet age, growing parallel to the digital technology that was supposed to supplant notebooks (the PalmPilot digital planner came out the same year as Moleskine's first notebook). Not only did the Moleskine notebook succeed in the face of disruptive digital competition, it situated itself as the ideal companion to smartphones, tablets, virtual note management services, and digital illustration software. It grew so successful that it has changed the behavior of a generation that was supposed to eschew handwriting into one where the paper notebook is omnipresent.

Moleskine today is a profitable, publicly traded company worth several hundred million euros, with annual sales of over €100 million, seven-hundred-plus products sold in over a hundred different countries, and more than two hundred employees spread between global offices and its rather anonymous headquarters in Milan (tucked into a courtyard, with no sign on the street). At the heart of all this is Maria Sebregondi, a woman an Italian newspaper once called "Mama Moleskine." Although she carries the somewhat innocuous title of VP of brand equity and communications, Sebregondi is the soul of Moleskine. I first met Sebregondi at the start of Design Week, in her sunny office. Dressed in a bright pink dress and purple glasses, in a uniquely Milanese way that's effortlessly elegant and imaginative, the silvery blond grandmother in her midsixties told me how a design career led to the creation of the familiar notebook on the desk in front of us.

Sebregondi was born in Rome; her father was an economist and her mother ran an editorial and graphics studio. After studying sociology, Sebregondi worked in publishing as a designer, wrote for design magazines, and set up her own studio in Milan, teaching creative thinking at the intersection of design, sociology, and trends. "My focus to design was the kinesthetic approach," she said, describing a method that emphasizes sensorial engagement. "We as human beings need to be stimulated with our senses, very physically. With sight, smell, taste, touch, and sound." When computers first captivated the design world in the 1980s, Sebregondi observed designers increasingly seduced by dematerialized, exclusively visual experiences. Over the long term, these left people wanting something more tangible. "Over [the past] thirty years that [digital dream] became a reality. But we discovered it wasn't only a wonderful thing. We really need physical objects and experiences."

During the summer of 1995, Sebregondi was sailing off the coast of Tunisia on the yacht of her friend Fabio Rosciglione. He consulted with the distribution company Modo & Modo, owned by another friend, Francesco Franceschi, which distributed design items and T-shirts around Italy. One night over dinner, under a sky bursting with stars, Franceschi started to talk about what kind of products Modo & Modo could manufacture on its own, rather than importing the designs of others. The conversation shifted to a question about who would buy those goods, and then to the changing nature of the world, which had just emerged from the cold war into the heady dawn of globalization. International travel was not only less restricted but more accessible, thanks to low-cost airlines. Technology, including inexpensive cellular phones, websites, and e-mail, allowed independent thinkers to become entrepreneurs and pursue their dreams unbound by geography. Speaking late into the night, the three realized that a new global creative class was emerging, driven by curiosity and passion. Sebregondi proposed that Modo & Modo create a toolkit for this individual, whom she labeled a "Contemporary Nomad."

Back in Italy, Sebregondi thought about what this nomad's kit would hold. There would be a great bag, a versatile T-shirt, the perfect pen, and maybe a utility knife. At the time, she was reading the book *The Songlines* by British travel writer Bruce Chatwin, an embodiment of her prototypical consumer. In one of the book's essays, Chatwin wrote about his preferred notebooks, which he bought in a particular stationery shop in Paris. "In France, these notebooks are known as *carnets moleskines*," Chatwin wrote, "'moleskine,' in this case, being its black oilcloth binding." The last time he returned to Paris, Chatwin discovered, to his great horror, that the family firm in Tours that had made his beloved notebooks was now out of business and the *carnets moleskines* were no more.

Chatwin's description of the notebooks, and the store where he bought them, struck a familiar chord with Sebregondi. She dug through boxes from her graduate studies in Paris nearly twenty years before, and pulled out her old notebooks, which matched Chatwin's exact description. *The Songlines* had been published in 1987, two years before the author's death, but Sebregondi held out hope that Chatwin's account of the notebook's demise proved false. After several inquiries to stationery stores in Paris, she confirmed that, yes, the firm had indeed gone out of business, and no one made that style of notebook anymore. Still, Sebregondi couldn't let the idea go. Soon after, she went to a Henri Matisse exhibit in Rome, and noticed that the artist's notebook matched the ones she had from Paris. Same with sketchbooks she saw at the Picasso museum, and in a photograph of Ernest Hemingway's desk. They all seemed to come from the same defunct French company. Sebregondi realized that the first product in Modo & Modo's nomad kit should be this lost notebook.

"This was something that could be re-created in a most refined way," she said. Over the next two years, Sebregondi and Modo & Modo worked to redesign, manufacture, and distribute the "mole-skine" notebook, which they positioned as a travel journal. Although there were many fine Italian paper manufacturers, in the end they settled on a supplier in China that was able to combine handmade details (tight binding on the spine, a hand-stitched pocket in the rear, perfectly flat seams) with the scale and cost needed for mass-market distribution. Their goal was to sell notebooks where notebooks had never been sold.

"The notebook market then was a completely unbranded market," Sebregondi recalled. There were cheap school notebooks, as well as fine handmade notebooks costing hundreds of dollars at stationery stores, but all were essentially nameless commodities. The only recognizable brand name at the time was the office organizer Filofax, which had already seen declining sales, thanks to computer calendars and other digital organizers. "The Filofax was strongly related to productivity and functionality," Sebregondi said. "If those are your focus, technology will kill you every time. That's why we went with imagination, image, and the arts."

The first branded Moleskine notebook saw an initial print run of three thousand, hitting store shelves in Italy and a few select European cities in 1997. Initially, the company refused to distribute to stationery stores. Instead Moleskines were sold in display racks at the cash registers of modern bookstores and design shops. It was presented as a "book yet to be written," one buyers were invited to fill with their own stories. The product quickly sold well among a small niche of writers, travelers, and the other global bohemians Sebregondi envisioned. Moleskine's market presence grew across Europe and into North America (I bought my first one at an art supply store in Toronto in 2005), but Sebregondi remained a "very part-time" contract employee at Modo & Modo, and mostly focused on other work.

The heart of Moleskine's transformation from a paper product into an analog cultural icon lies not simply in its artful design—soft, creamy paper that practically invites a pen's ink; rounded corners that ease the notebook into a pocket; a cover that's hard enough to keep pages from bending, but soft and almost leathery in feel—but in the myth that Sebregondi wrapped it up in. Through the packaging, and a story that folded out from the inside cover, the Moleskine (a name trademarked by Modo & Modo in 2006) was presented as "the Legendary Notebook of Hemingway, Picasso and Chatwin" with tales about its place at the core of modernity's greatest art and literature. Whether the new Moleskine notebooks were actually the same ones used by these canonical artists (they weren't) was not specifically the point.

That foundational myth, which Moleskine continues to hammer home in all its press releases, marketing materials, and interviews, is essential to understanding the emotional power of a resurgent analog brand. From the get-go, the company knew that its notebooks wouldn't exclusively contain the brilliant creations of the next Picasso. There would be a lot of melodramatic teenaged diaries, half-baked doodles and class notes, and grocery lists. But because they were written in a Moleskine, they would still *feel* more creative than if they were scribbled on another piece of paper.

"*Creativity* is a word that's now completely sold," Sebregondi said, "but the concept behind it is strong and real. People want to be creative and feel creative, even if they are not. Creatives have the ability to create an emotional trigger, and the analog world is the one able to create this emotional attraction and experience."

This formed an almost tribal identity around the Moleskine notebook and those who used it. The notebook became a symbol of aspirational creativity, a product that not only worked well as a functional tool, but that told a story about you, even if you never wrote on a single page. Like a Patagonia jacket or a Toyota Prius, it projected someone's values, interests, and dreams, even if those were divorced from the reality of their lives. This is why Moleskine never needed to advertise, and never does to this day. Each notebook spotted at a coffee shop table, or in the hands of a journalist, was worth more than any billboard or magazine page. "This is a company that went from being a category maker to a category icon," said Antonio Marazza, general manager at the Milan office of the global branding agency Landor Associates. "The emotional and aspirational capital Moleskine can deliver goes beyond stationery." It buys you access to the most select group of consumers, the cool creatives who set trends that other consumers invariably follow. Republican strategist Karl Rove once cited Moleskines as a signal flare of liberal pretentiousness. Marazza believes that this is only possible because Moleskine is a physical, tactile product. "It is easier to tell an interesting story about something physical than something immaterial," he said.

Branding is nothing unique to Moleskine, but a great branding campaign is only one half of the equation. The other indeed goes back to those qualitative attributes of the paper notebook itself. "This notebook," Sebregondi said, holding up a classic black Moleskine, "is a physical experience that can leave space to the imagination. That is greater than technology." Creativity and innovation are driven by imagination, and imagination withers when it is standardized, which is exactly what digital technology requires—codifying everything into 1's and 0's, within the accepted limits of software. The Moleskine notebook's simple, unobtrusive design makes it feel like a natural extension of the body. It doesn't interfere in your personal style, and because of this, it allows for an undiluted physical recording of your mood. "All that is lost when you standardize," said Sebregondi.

Similar to the analog studio recordings in the last chapter, the physical constraints of a blank page present a certain creative freedom. Antonio Marazza told me a story from a decade back, when all the designers at his firm first received Adobe's Photoshop software. Overnight, the quality of their designs seemed to decline. After a few months of this, Landor's Milan office gave all their designers Moleskine notebooks, and banned the use of Photoshop during the first week's work on a project. The idea was to let their initial ideas freely blossom on paper, without the inherent bias of the software, before transferring them to the computer later for fine-tuning. It was so successful, this policy remains in place today.

Such practical use drove Moleskine's significant early growth, as a whole new type of user, who was very much the opposite of Sebregondi's dreamy, type B nomad, embraced the notebooks. These were computer scientists, high-achieving executives, and type A experts in productivity and efficiency. They took to the Moleskine not out of any sense of adventure or an affinity with Hemingway, but as the best, most efficient tool to organize their thoughts. Followers of David Allen's popular time-management method, Getting Things Done (which began as a best-selling book in 2002 and was branded a "New Cult for the Info Age" by *Wired*), adopted the Moleskine notebook as their preferred tool, transforming an object designed for romantic, creative scribbling into a hammer of productivity with charts, lists, and bullet points. The Moleskine had no special features to invite this, aside from its simple design. It just happened to be the perfect blank slate for Getting Things Done devotees to hack for their purposes, a true tabula rasa. "Getting Things Done isn't a paper-dependent method," Allen told me last year. But, he said, the "easiest and most ubiquitous way to get stuff out of your head is pen and paper." A pen and paper requires no power source, no boot-up time, no program-specific formatting, no syncing to external drives and the cloud. "You can waste time with all kinds of stuff," Allen said, "but the digital world provides a lot of opportunity to waste a lot of time." And wasted time was the stated enemy of Getting Things Done.

In his book *The Organized Mind*, cognitive psychologist and neuroscientist Daniel Levitin talks about the tremendous harm inflicted on us by information overload, which he claims is worse for your brain than exhaustion and smoking marijuana (he calls multitasking "empty-calories brain candy"). Numerous studies have shown that handwriting notes is simply better for engagement, information retention, and mental health than is writing on digital devices. "Writing something down conserves the mental energy of worrying that you might forget something and in trying not to forget it," Levitin wrote, presumably on paper.

Paper holds other advantages as well. "I can't tell you how much money I've made as a tech consultant helping people recover files saved in formats that no longer exist," said Patrick Rhone, a computer consultant in Minnesota who runs the pen blog *The Cramped*. "That'll never happen with my paper notebooks. Those will be able to be read for the next ten thousand years, if they still exist. People know that this thing they hold in their hand can never be taken away, except for fire, disaster, or if it is physically removed. It can live for eons. That beats any backup any digital file can promise me." I learned this the hard way, back in university, when an entire semester of notes I took on a small laptop failed to sync to my desktop computer, wiping out three months of lectures. Only photocopies of handwritten notes from my friends saved me from failing those courses.

The final factor that unleashed millions of Moleskine notebooks on the world (close to 20 million last year, actually) was the convergence of these productivity users and the Internet. Around 2002, blogs dedicated to Moleskine notebooks sprang up, including *Moleskinerie* (a visual gallery of Moleskines filled with beautiful drawings), *Getting Things Done*, student blogs from computer science classes at MIT, and Mike Rohde's sketchnote community.

Rohde, a web designer, cracked open his first Moleskine notebook in 2007, at a conference in Chicago. "The physical book provided me with limitations, and that pushed me to be creative," said Rohde, who usually took verbatim notes on a laptop. "A limited page size meant I couldn't take the detailed notes I used to. I used visual images to make the notes more structural and fun to do." Rohde's sketchnote method, which combines illustrations with writing to effectively communicate ideas, is so popular that he has authored two best-selling books, and there are several online communities devoted to the practice, where tens of thousands of users share photos, illustrations, and videos of sketchnote-filled Moleskines.

———◆———

"Moleskine benefitted heavily from the web," said Marco Ariello, the chair of Moleskine's board, and a partner at Syntegra Capital, the private equity firm that purchased Moleskine in 2007, who spoke with me in his Milan office. "Via the web, you could really see the engagement of the Moleskiner. It was a passion that went beyond the functionality of a notebook." Those online communities hinted at a much larger market potential for the company, and a loyal bulwark against competitors.

Syntegra's acquisition valued Moleskine at €62 million, over four times the company's annual sales at the time. Skeptics abounded. Ariello recalls one newspaper joking that Syntegra had paid Internet valuations for a paper company, hardly a rapid-growth industry. How could Syntegra grow a business built by selling a $20 notebook, when similarly sized paper notebooks retailed at one quarter the price? But Syntegra saw something deeper in the company beyond the notebook, and that was the long-term potential of the Moleskine brand.

That brand, which Sebregondi was brought on full-time to develop, is crucial to understanding the success of not just Moleskine, but where analog is capable of expanding beyond a small niche. "I can't find any more compelling reason for this product beyond *authenticity*," said Prof. Andrea Ordanini, chair of the marketing department at Milan's Bocconi University. That authenticity comes from the implicit freedom in a blank Moleskine, a supply of blank pages with the unlimited potential for creativity. "Even technical people using Moleskines, like the Getting Things Done folks, are specifically the people who need to feel free," said Ordanini. The message a Moleskine sent, and which the company is building a broader brand around, is the notebook as aspirational capital. You can be a buttoned-up investment banker, or a nerdy engineer, but open up your Moleskine and you are playing on the same creative field as the cool architect designing an urban park at the café.

"What Moleskine does, by allowing you to differentiate yourself via a product, design, and premium pricing, is show that you are different than others," said Prof. Carlo Alberto Carnevale Maffé, who teaches strategy and entrepreneurship at Bocconi's business school. A big part of this, Maffé told me, is Moleskine's status as an analog product. When everyone wields a phone made by either Apple or Samsung, a paper notebook stands out. While such services as Twitter and Facebook are "macrosocial" networks, engaging vast swaths of the global population in largely impersonal relationships, niche analog brands, such as Moleskine, create highly personal "microsocial" networks. The act of buying a Moleskine notebook is costly and exclusive enough that you might feel an affinity with a stranger who pulls out the same one you own. Twitter and Facebook, by contrast, are both free and so ubiquitous that the sight of someone checking an account in public is unremarkable, or even slightly annoying.

Then there is the question of legacy. In digital, a legacy brand is yesterday's lunch, because the best digital technology is always the next one, and consumers have no loyalty to the past. In analog, legacy commands a premium. "You can perfectly imitate a Louis Vuitton bag," Maffé said, pointing at one hanging on the arm of an elegantly dressed woman sitting near us in the luxury hotel café where we were having coffee, "but you can't sell it for six thousand dollars because you don't have the heritage. As long as you can convince users the past is relevant, they'll pay billions for it. There is no rational economic reason behind that. Just marketing." By sticking to its brand image as the legendary notebook of modernity's greatest minds, Moleskine continues to enforce its legacy status, even if it is largely contrived.

———————◆———————

I returned to Moleskine's offices on my second morning in Milan, to speak about the company's brand with Arrigo Berni, Moleskine's CEO. While Sebregondi's office is a visual riot of color, his is muted and minimally furnished. Berni, who has a shaved head and a closely cropped white beard, came to Moleskine through Syntegra's acquisition. His experience included household packaged goods at Procter & Gamble, management consulting, and luxury goods, first with Bulgari and later with the leather company Testoni. When he first began looking at Moleskine, Berni was drawn to how consumers spoke about the brand. They cited an emotional bond that he had never seen before, even with Bulgari diamond necklaces worth tens of thousands of dollars, and he found it ironic that this quintessentially analog product's success was so tied to its digital fans.

"If this had begun in the 1980s, it wouldn't have flown," Berni said. "The idea of analog, besides the physicality of the product, is the experience made possible between a human being and that object. . . . The analog businesses succeeding today, despite digital competition, are those that have been able to really create and emphasize *credibility* . . . the experiential dimension of what they sell. That may seem like a game of words, but it defines the difference between a functional product," such as competing notebooks from the French paper company Rhodia, which mainly advertises its physical features, "and a tool for experience. This way of looking at the business is part of the brand's DNA from day one."

I found my conversations with Berni and Sebregondi fascinating, but in a way, also frustrating. As happy as they were to speak about the intangible nature of the Moleskine brand, and how the spirit of Chatwin drove consumers to it, whenever I asked about the practical, functional elements of a Moleskine notebook, and how *those* fit into the brand, they bobbed and weaved like Hemingway in a boxing ring. There clearly was a directive to stay on-message, and focus on the bigger Moleskine brand, rather than the notebooks that still drove the bulk of sales (over 91 percent of Moleskine's revenue in 2014 came from paper products). Speak about elastic brand identity all you want, but at the end of the day Moleskine is a company that makes the most recognizable notebooks in the world. Why shy away from that?

I partly got a sense from Berni's answers to another question, about the company's 2013 debut on the Italian stock exchange. In the six years between Syntegra's acquisition of Moleskine and its IPO, the company had achieved steady growth, expanding its line of notebooks into many different colors and sizes, direct-to-business sales (a huge market is to technology companies, such as Facebook), and special editions, from wine and food journals to ones featuring LEGO and Star Wars characters. Sebregondi had begun building out the original toolkit at the root of the business, introducing Moleskine bags and iPad cases that resembled the notebooks, as well as pens, and even reading glasses. With Moleskine products selling in over twenty thousand retail locations worldwide, the company had achieved a truly global presence. In the IPO prospectus, the investment bankers at Milan's Mediobanca noted that the brand's unique positioning, "which expresses a set of intangible values (namely culture, design, imagination, memory and travel) and is endowed with a unique heritage," was one of Moleskine's biggest strengths. There was ample growth potential for new and emerging markets, room for brand extensions into other products, a chance to develop Moleskine's own retail stores, and the potential of collaboration with digital partners. Revenue and net income had all grown steadily, while costs and debt were decreasing.

Moleskine's stock made its debut on the Milan stock exchange on April 3, 2013, at a price of €2.3 a share, valuing the company at €490 million. It was the first IPO in Italy in a year, and a particularly proud one for a country that had been hit hard by the eurozone recession. The IPO would have been a triumphant success for any company, and an especially sweet note for analog's revenge, but Moleskine shares almost immediately took a sharp turn south. Despite consistent growth, profits, expansion, and other markers of financial performance, Moleskine's stock price sank through the rest of 2013 and 2014, bottoming out at year's end, at just €1 per share. Things began to improve in 2015. By last April, the stock finally cracked the €2 mark. Still, Moleskine has a long way to go just to recoup the losses of investors who bought shares at its IPO.

Several factors contributed to the drop in the stock, Berni said. Moleskine was lumped together by analysts with other Italian luxury brands, such as Prada, Gucci, and Salvatore Ferragamo, which were primarily in the clothing business. Despite the rhetoric of intangibles and premium pricing compared with other notebooks, Moleskine is not a luxury brand, and operates very differently from those companies, whose products are seasonal and sell at vastly higher margins. Analyst projections were also overly optimistic; the Mediobanca IPO report estimated the company's sales and profits would nearly double every two years. But Berni also felt the company's stock was a victim of prejudices around paper, and analog businesses in general.

"I sat in meetings with bankers and investors, and they pointed to the iPad in front of them and told me, 'In three years you are going to be out of business. There is no future in paper. Look at my iPad! People will stop writing.'" Berni said. "There was no way to get them not to make the mistake of projecting themselves onto the market. People were starting out with the belief in paper's demise." Many of those same skeptics took up short-selling positions, essentially betting against Moleskine's stock, further driving down the share price. "There will always be a certain part of the world who will never be able to reconcile our success," Berni said, adding with a wave of his hand that "these are the same people who say 'But we live in a digital world!'"

What the skeptics failed to see—beyond the surge of new apps, social networks, and connected digital devices—was that Moleskine's greatest achievement was making the art of note taking a key behavior of the digital era. Whether it was a journal entry on vacation, a brainstorm for a new startup, such as AirBnB, or just the daily scribbling of work and life, Moleskine had made the action of putting pen to paper desirable for tech-savvy consumers, and others saw an opportunity to profit.

The market is now swimming in notebooks. I poked around a bookstore one afternoon in Milan, and saw no fewer than two dozen notebooks that looked exactly like Moleskines—from the dimensions to the curved corners to the elastic band—but none were made by Moleskine. There are Moleskine-looking notebooks made by older paper companies such as Lechtturm 1917, and generic ones churned out by the boatload overseas, with any number of designs (a sports team's mascot, a museum logo, Mickey Mouse) slapped on the cover. There are high-tech ones, such as Mod Notebooks, which you send back to the company and have scanned into the cloud for you, and a world of Kickstarter-funded notebooks, launched by people who believe they can improve on Moleskine. The Spark Notebook was created by Kate Matsudaira, a tech executive at companies that included Microsoft and Amazon, and is a goal-oriented notebook/planner "designed for people serious about success." Baron Fig's notebooks are geared toward "brainstorming and ideation," according to the company's cofounder Joey Cofone. If you want to take notes and feel good, then opt for notebooks by Public-Supply, which donates a quarter of each sale to public school classroom supplies.

Most of these notebooks look and feel remarkably similar to a Moleskine notebook. Brand seems to be the major (and perhaps only) differentiator among these companies, and the most successful notebook brand to emerge, post-Moleskine, has been Field Notes, which bases its notebook designs on old American agricultural journals from the 1930s. The Field Notes product is "deliciously analog," according to brand manager Michele Seiler. The notebooks are hand-made in the USA, and printed with all sorts of quirky facts and fig-ures inside the covers. Each year the company puts out a series of limited-edition notebooks (State Fairs, US Polar Expeditions), which have become collectors' items, resold for several times their price on-line. Field Notes are mainly sold at clothing stores, bike shops, hard-ware stores, gun ranges, and other retailers big (Nordstrom) and small (barbershops).

People don't jump into starting a notebook company out of nos-talgia or romance for the golden age of writing. They do so because there is a viable, growing consumer demand for these objects, and a market that values paper. Not *all* paper, and certainly not the lowest-priced, commodity paper. No one is going to send a fax when the same document can be e-mailed, and many people prefer electronic bills to paper bills. But what Moleskine and its competitors demonstrate is the tremendous potential for innovation and profit in a previously dis-missed analog industry, and the ability of branding to create paper products that people will pay for.

By losing its job as the dominant form of communication, paper has been elevated to an exalted place, where it can play off its intan-gible analog advantages. In the process, it became cool, in the way that candles and bicycles are cool, even if they are technically "obso-lete." New letterpress printers and stationery companies are popping up in every city, and one of the top-selling categories in publishing is adult coloring books. New pen, stationery, and paper boutiques are opening around the world, including one store dedicated exclusively to pencils that opened in New York in 2015. Digitally enabled paper businesses, such as Punkpost (which will handwrite and mail a card for you) are also growing quickly.

The success of paper has even convinced some digital businesses to embrace analog. "In hindsight, maybe this wasn't the best name for us," conceded James Hirschfeld, CEO of Paperless Post, a premium online invitation company that began in 2009. Following unrelenting requests from its customers, Paperless Post launched a product line in 2012 called Paper by Paperless Post, which is really just a line of paper invitations. "Obviously it's an oxymoron," Hirschfeld told me, when we spoke at the company's office in New York. "Our customer really lives a hybrid life, communicating between paper and digital devices," Hirschfeld said. "In order to stay relevant, we had to offer a physical product to customers." It has paid off. More than half of Paperless Post's new business now comes from paper.

Paper's revenge isn't exclusively relegated to emotionally tied consumer goods, such as cards and stationery. As Moleskine's adoption by various professionals shows, there is a lasting utility in paper. The perfect case for this is the British company MOO, which began in 2004 under the name Pleasure Cards. The purpose of a Pleasure Card was the web code printed on it, which allowed you to access an online profile on the company's social network. It never took off, but people really liked the design of the cards, and requested them as regular business cards.

"The simplest idea prevailed," MOO's founder and CEO Richard Moross told me from his office in London's East End, dressed in a sea-foam mint blazer and coral trousers. "Here was this five-hundred-year-old piece of technology, bigger than e-mail, mobile, or anything else. It required no batteries or software, and yet it was completely global. You can give one of these to anyone anywhere in the world," he said, handing me his card, "and they will know exactly what it means."

MOO uses digital technology to its full potential, allowing for completely customizable designs and printing options, while pushing the visual boundary on business card design. With MOO's Printfinity cards, for example, you can order a set of cards with a completely different image on the back of each card. With its Luxe stock, the card is a three-layer sandwich of different papers and colors. Each March, Moross attends the technology conference South by Southwest Interactive in Austin, Texas, where he hears about the next app predicted to kill the business card. Meanwhile, everyone in the room is still ordering more cards from MOO, which supplies stationery to such technology companies as Skype, LinkedIn, and Uber. "Attempts to reinvent business cards for the digital age have got nowhere," *The Economist* wrote in 2015. "Even at the trendiest of Silicon Valley tech gatherings, people still greet each other by handing out little rectangles made from dead trees rather than tapping their phones together."

The reason why MOO business cards succeed, even in the most digitally driven industries, is that paper cuts through the noise. Moross gets thousands of e-mails a day, and deletes most of them unseen, but he opens every single envelope that arrives on his desk. Even Google and Facebook distributed elaborate printed publications to political and business leaders, because they stand out. A beautiful paper object has a far greater chance getting past the gatekeepers of assistants and into the hands of such people as Michelle Obama and Bill Gates than does a quickly deleted attachment. Yes, a business card is an anachronistic tool for storing contact data in an age of synced customer management databases and easily searched LinkedIn profiles, but it still makes that crucial first impression a lasting one, and leads to future impressions every time you lay eyes, or fingers, on that card. That effectiveness cuts across industries and generations.

"Digital natives are actually the most interested in paper," said Chris Harrold, the creative director at Mohawk Paper, an eighty-five-year-old paper company in upstate New York that supplies MOO. "They don't have a nostalgic association with it. They find it really beautiful, and refreshing. Their digital devices are a commodity; a commodity delivery platform. But print has an ability to organize information in a special way. The web is just this endless loop of information."

———◆———

*T*he web's endless loop may be Moleskine's final act of revenge, as Moleskine's core product—paper notebooks—become increasingly integrated with a host of digital companies. The first, which began in 2012, was a partnership with the cloud-based note-taking service Evernote, and involved the creation of a Moleskine notebook with special paper that made scanning handwriting with Evernote's smartphone app much more accurate.

When Evernote first approached Moleskine, Sebregondi and others in Milan were initially suspicious. After all, Evernote was a tech firm that had made the paperless office its explicit goal. "When we started the relationship with Moleskine, we declared a truce with paper," Evernote's VP of design, Jeff Zwerner, told me. "We poked fun at [paper], but it's not an either/or. We are realists that we live in a world of multifaceted communication." The downside of digital technology, Zwerner said, is that it is constantly changing. As soon as you understand and figure out how to do something, the next version of hardware or software comes along, and the learning process starts fresh. "With this," he said, grabbing the Evernote/Moleskine Smart Notebook in front of him, "you understand. You pick it up and know how it *works*."

Moleskine has entered into other partnerships with digital services, producing notebooks that work with the handwriting transcription pen Livescribe, photo albums with the digital printer MILK, and custom books with Paper FiftyThree. "Technology has created wonderful opportunities to make these physical experiences possible," Sebregondi said, as we walked through a showcase of young Belgian designers in the Lambrate district of the city. The company's first forays into digital were entirely accidental and organic: the blog posts of early users, who shared their drawings or organizational systems online, and created this global community around a simple paper notebook.

We entered an open industrial warehouse that was filled with designers and products from around the world, and went to the Moleskine pavilion in the back, which was on a raised platform surrounded by a riveted metal frame. Here, the company was demonstrating its latest digital collaboration with the software giant Adobe, and the integration of Moleskine notebooks with Adobe's Creative Cloud service. It worked similar to the Evernote notebook, with specially marked pages that allowed an Adobe app to accurately scan drawings and text made in the notebook into a variety of file formats. Throughout Milan Design Week, a number of designers had been given these notebooks, and the Moleskine digital team was uploading their drawings onto the Adobe software, letting the designers edit them, and printing out the results to display. It seemed to work well, from a functional perspective, but more important, these digital collaborations made sense for Moleskine because they put analog first, presenting paper as a desirable creative technology, then finding new applications for what it does best.

"We are very physical creatures," Sebregondi said, examining a drawing of a furniture exhibit one designer had just scanned from his notebook, "but we are also very egotistic. We need to show and share."

Still, there is a delicate dance at play here. Many successful analog companies have wasted untold money and time chasing the siren of digital innovation over the past decades, and the market is littered with the remains of those who felt they could transform stable analog firms into high-tech wonders. Moleskine has had modest success with digital collaborations, but it cannot forget, despite all the talk of intangible branding and potential extensions into other products and services, that it is a company whose success, growth, profits, and identity come from selling paper notebooks. I can understand why Sebregondi, Berni, and others at Moleskine are cautious about projecting that too eagerly, but paper is Moleskine's greatest strength and the reality of its existence.

The market, which once punished Moleskine for its paper heart, might just agree. Over coffee across from Milan's famous La Scala opera house, I spoke with Alberto Francese, a research analyst with the bank Intesa Sanpaolo, who had recently begun covering Moleskine's stock. "At the end of the day it's a paper product," Francese said, noting that while Moleskine's collaborations and brand extensions were all positive, the company's main driver of growth would be increased sales of notebooks to newly educated and urban consumers in such countries as China and India, and in Latin America. (In early 2017, Moleskine was fully acquired by a Belgian luxury-car distributor for just over half a billion euros.) "Is it desirable to separate Moleskine the brand from the notebook?" asked Leonardo Proni, a lawyer in Milan who worked on the company's IPO. "No way." Even though Moleskine's emotional brand went well beyond its notebooks, such a move would "be complete suicide."

As Sebregondi and I walked through more design exhibits that warm spring evening, I continued to press her about the company's analog future. I could tell she was tired of my questions, but I needed to know whether Moleskine, arguably the poster child for the Revenge of Analog, was still committed to paper. The company's research division was working on new paper projects, Sebregondi conceded, hinting at new designs for Moleskine notebooks and other paper products that would come out in the next few years. What mattered, Sebregondi stressed, was Moleskine continuing to create products that provided its customers with a vessel for wrestling down the ideas banging around their heads into a tactile, permanent record.

Later, when I was going through my notes back home, I read over one of Sebregondi's final comments to me that night. "This company is fostered all the time with new stories, talent, and exploration," she said. At the time I took her remark as one more nod to the supposedly legendary notebook narrative everyone at the company kept pressing on me, but then I looked up, past my computer screen, at a bookshelf on the far side of my office. There, lined up in neat rows, were dozens of notebooks, including several Moleskines, that held every interview and note I had taken as a journalist over the previous thirteen years. These notes were at the core of the three books I had written, notes behind hundreds of magazine and newspaper articles, countless random observations, sketches, and conversations from all over the world. They stood there on those shelves as permanent testaments to my own stories, talent, and exploration. As Bruce Chatwin wrote in his fateful essay that inspired Sebregondi twenty years ago:

"I felt, before the malaise of settlement crept over me, that I should reopen those notebooks. I should set down on paper a résumé of the ideas, quotations and encounters which had amused and obsessed me; and which I hoped would shed light on what is, for me, the question of questions: the nature of human restlessness."

3

The Revenge of Film

*M*oleskine's *story demonstrated how an* analog company could forge a new future for old products, but paper was still a relatively healthy business. What was the way forward for an analog industry that had truly been decimated by digital, and whose very complexity was its disadvantage? From Milan, I rented a car and drove south to find out. Flat plains gave way to steep hills, sunshine turned to rain, and in less than two hours the silver expanse of the Mediterranean appeared. I drove west along the coast, then back inland, passing through a series of long, dark tunnels into rolling misty mountains. As the road brought me deeper into Liguria province, I kept passing underneath what looked like an old ski lift with heavy iron buckets hanging from its cables, slowly ferrying its cargo from the sea to some unknown destination. Finally, I arrived in the valley simply known as Ferrania.

Ferrania is not so much a place as what remains of one: the ruins of a huge industrial village, which at one time represented the bright light of Italy's manufacturing sector. Although the company Ferrania Technologies still produced chemicals for the pharmaceutical industry here, and a relatively new (though largely inactive) solar panel plant stood out among the decaying, abandoned buildings, the place was a ghost town. Wind blew through the shattered windows of offices and apartment buildings left to rot in the damp air. Weeds reached the tops of goalposts on the soccer pitch. The only sounds, aside from the occasional truck gearing down, was the echo of steel and concrete twisting and crunching as a demolition crew slowly disassembled the buildings. Above the valley, the mysterious industrial gondola steadily carried its cargo, which I later learned was coal offloaded from ships and hauled over the mountains to the nearby power plant at Cairo Montenotte, a black smoky place straight out of Dickensian England. Ferrania is far from the picturesque Italy of *La Dolce Vita*. It feels more post-Soviet than anything.

I drove past an unmanned security gate, down a heavily rutted road, and underneath corroding pipelines, before arriving at a four-story building in the far corner of the valley. It had yellowed windows, and its facade was a drab mix of faux midcentury brick and patched concrete that was peeling like old skin. By the small door there was a sign cast in concrete that said this was the "laboratory recherché fotografiche."

A small Land Rover came down the road in a plume of dust, and out jumped Nicola Baldini in a brown waxed jacket, tinted Armani glasses, and green Doc Martin boots.

"So," Baldini said, opening the door to the building they call the LRF, "are you ready to see the future of film?"

Tall, with a graying beard and sweeping gestures, Baldini thus launched into a passionate mixture of chemistry lesson, industrial history, and business pitch that would continue for the next twenty-four hours, stopping only for sleep.

What Baldini and his business partner Marco Pagni were attempting to do here, at this small, crumbling factory in this remote, bankrupt valley, was revive the production of new color still and motion picture film under the old FILM Ferrania brand. To do that, they had to overcome every single obstacle facing an analog industry in a postdigital economy. To get FILM Ferrania to the point where it could make even a single roll of film, Pagni and Baldini needed to rescale an assembly line designed for mass-market production to a tenth of its size, with a skeleton crew, patchy knowledge, dangerous and discontinued materials, and a bare-bones budget. They had to finance and engineer the relocation of huge machines from the original film factory buildings nearby to this smaller facility, all in the few months before the demolition crew tore down those other buildings. If they missed this narrow window, the chance to rebuild FILM Ferrania would be lost forever.

This is a good reminder. Analog may be having its revenge, but it still takes courage—a lot of it—to start a business like this in the digital era. It is one thing to dream about making film or records, or reviving other analog products, but turning those dreams into a physical reality means coming face-to-face with the tremendous disadvantages of industrial analog production, and of selling analog goods to consumers. Scarce materials and knowledge, diminished economies of scale, outdated machines, deadly chemicals, unstable markets, variable quality, and endless unforeseen, money-sucking headaches await. Pull this off, and you are rewarded with a product, such as photographic film, that is temperamental at the best of times and all too often completely dysfunctional. Witness even a slice of this (as I did in Ferrania), and you quickly appreciate why the move to digital photography was so swift and widely embraced. This is the industrial equivalent of raising a sunken ship from the seafloor and *then* refurbishing it to sail again.

---

*F*ittingly, Ferrania's origin is volatility. In 1882, an explosives factory (called SIPE) was set up in the valley, which produced bombs and shells for Russia's imperial navy. The emergence of the Soviet Union in 1917 cost SIPE its biggest customer, and the firm shifted to photographic film. The company was renamed with the acronym FILM (Fabbrica Italiana Lamine Milano) Ferrania, and began producing still and motion picture film in a variety of formats. During Mussolini's fascist reign, Ferrania film was the obligatory option for Italian photographers and directors, but after the war, it was adopted by the vanguard of the country's filmmakers, such as Federico Fellini, whose look was shaped by Ferrania's proprietary chemistry, which captured the curves of Sophia Loren and the smirk of Marcello Mastroianni in shades and colors that became recognizably Italian.

For most of the twentieth century, if you took a photograph or made a movie in Italy, it was most likely on film made in Ferrania.

The brand was Italy's equivalent of Kodak or Polaroid—not just a film product, but the repository of memory associated with birthdays, vacations, and milestones. At its peak, Ferrania employed four thousand workers. It wasn't just a company, it was a fully enclosed community, a valley whose people and resources existed to make film. You can see photographs from this heyday online: packed lunch halls, women working under shiny microscopes in pristine laboratories, company beauty pageants, and football tournaments.

FILM Ferrania's fate changed in 1964, when Ferrania was purchased by America's 3M Corporation. Over the years, 3M used the Ferrania brand less and less, and though manufacturing was still based in the valley, most of the film was sold under different names, such as 3M Scotch, Imation, and Solaris, or simply the name of a retailer (at one point, Ferrania was the largest supplier of private-label film worldwide). Still, Ferrania was a model factory with a corporate reputation on par nationally with Fiat, a center for imaging and chemical innovation used by various 3M businesses and competitors, such as Panasonic and Kodak, which wished to tap its unique expertise. Ferrania's campus sprawled over nearly 550,000 square yards across twenty buildings. The core of this was the LRF building that 3M built in 1967 (the very one where I met Baldini and Pagni), which was the research and development headquarters for the film group. Those who worked in the LRF were the best and brightest in the company, the "top of the top," in Baldini's words. "These men were gods in this valley."

The end of Ferrania mirrors that of the wider film business. 3M restructured the company in 1995 and first sold off its lucrative medical imaging wing, then unloaded the commercial film division in 1999 to an Italian investment company. That was the last year of film's ascendency, the peak of global production. Kodak had created the technology behind digital photography years before, but by the early twenty-first century, image sensors, computer processing power, and storage capacity had come down sufficiently in price to make consumer digital cameras viable competitors to film. Most people took one shot with digital and never looked back.

I remember my own transition with 5.0-megapixel clarity. I used to be into photography as a teenager, wandering around the city with my friends and a fully manual Pentax K100 loaded with Ilford black-and-white film. We'd use the high school darkroom to develop and enlarge these shots, emerging from the red light smelling of chemicals with a single perfect print. When I was eighteen, I traded the Pentax for a Canon Rebel G, a film camera with digital sensors that automated focus and light adjustment. I still took the camera everywhere: shooting rolls on travels and ski trips, trying to replicate what I'd seen in such magazines as *National Geographic* or *Powder*. I filled dozens of books with these photographs. I had no real talent, but it was something I loved.

In 2003, I moved to Buenos Aires, Argentina, to get my start as a freelance journalist. Before I left, I picked up a Canon PowerShot digital camera, which I swore I would only use for work. I managed to shoot and develop one last roll of film in Argentina, but halfway through the second one, I was so enthralled by my digital camera—its instantaneousness, shareability, cost, and total freedom—that I simply forgot my film camera even existed. When I left Argentina two years later, I gave the untouched Canon Rebel to a girlfriend. "What should I do with the roll in here?" she asked me. I told her to throw it out.

Digital's advantages over film were so decisive, and came on so rapidly, that the industry had no chance to adapt. Film just dropped off a cliff. The factories were too large, and built for too high a volume, to handle that drastic a decrease in production. In the United States alone, 800 million rolls of film were produced in 1999. By 2011, that dropped to just 20 million rolls. Polaroid went bankrupt in 2002, Ferrania the year after, followed by Britain's Ilford and Germany's Agfa in 2005. Eastman Kodak filed Chapter 11 in 2012, and now has only eight thousand global employees, compared with over 145,000 at its peak. There are dramatic photographs of Kodak factories being dynamited into oblivion, with the telltale LCD screens of digital cameras and phones in the foreground capturing the destruction with telling irony. Kodak's still film production was spun off into a separate company, called Kodak Alaris, now owned by the company's UK pension fund, while Eastman Kodak in the United States stuck with motion picture film. Only Fujifilm escaped bankruptcy by diversifying its business away from imaging and investing early in digital cameras, but not without discontinuing its motion picture division and drastically cutting back on still film products.

Ferrania's film division endured successive rounds of layoffs until the LRF research building was simply locked up by management one weekend in 2006, and all its employees let go. The very last roll of film was produced at Ferrania in early 2011, when the last employee finally shut off the machines.

Baldini and Pagni told me this tale over a lunch of lasagne and braised rabbit in a small restaurant a few miles away. The two live in Florence, where Pagni (who speaks in a measured softness, and looks like a slender Russell Crowe dressed in baggy sweaters) owns a Kodak film-processing laboratory. He has been in the film business since he was fifteen, working initially as a movie projector technician. A self-taught engineer, Pagni has an encyclopedic knowledge of the entire film-manufacturing and -developing process. He is not a photographer, but did own a vinyl record press for a brief period. Baldini is a lifelong amateur filmmaker, camera collector, and photographer, but his day job is as a professor of computer science, specializing in artificial intelligence, machine learning, and big data.

In 2011, Baldini came to Pagni's lab (he had been a customer since 2006) with a large roll of 35 mm Fujifilm motion picture stock that he wanted to cut into smaller formats, such as 16 mm and 8 mm, so it could fit in smaller cameras. To do this they needed a special machine, called a perforator, which could precisely cut the film to the right size. "This is not exactly equipment you can buy from the supermarket," Pagni said; "you have to buy it from other companies." Fujifilm and Kodak were of no mind to sell two random Italians their proprietary machinery, and others, including Agfa, had already destroyed their perforators. Luckily a friend told the two that a new photography museum in Florence contained a perforator from FILM Ferrania. Baldini and Pagni went to the museum and saw the serial number 86 on the machine. "Okay," Baldini recalled asking Pagni, "so, where are the other eighty-five?"

At the end of 2011, the two drove to Ferrania, and knocked on the shuttered factory door to inquire about the machines. No answer. They poked around town, and finally got a former manager on the phone, who contacted the current stewards at Ferrania Technologies, who informed them that the buildings were slated to be demolished, but that they were free to take a look at the contents. Touring the mammoth film-production lines, they saw that everything was in pristine condition, and quickly located a perforator. Baldini bought it for €10,000, the price of scrap. The thing weighed a ton, and they arranged to return and figure out how to move it.

When they came back in early 2012, Pagni and Baldini had already located a potential site at which to set up the perforator—in Bulgaria, where the labor would be considerably cheaper. But as they toured Ferrania with Marco Descalzo, the last manager of the photo division (kept on to sell off Ferrania's assets), he took them to the LRF with a representative of the Liguria government, which now owned the research building and its property. Pagni and Baldini began to ask increasingly ambitious questions. What if they needed to package the film? Should they also buy a film-packaging machine? If Kodak, Fujifilm, and the remaining players in the film market were continuing to exit the business, how would they secure a steady supply of film to perforate? Why move these huge machines to another country, when everything was sitting right here in Ferrania, and the formulas were specifically calibrated for the nearby climate ("like a pizza dough," said Baldini)? Why not just make film here? Why not revive FILM Ferrania?

Descalzo told them they were completely crazy. "Ferrania is dead and not another meter of film is coming out of here!" The next day, he reluctantly told them that perhaps it might just be possible, for one very important reason. When the LRF building was set up by 3M, it was essentially designed as a miniaturized version of the larger factory next door. That meant that the largest, most complicated machinery in the production of film, the processes that were physically inseparable from the building, were not only intact within the LRF, but were the ideal size for the smaller production quantities that today's niche film market can sustain. To make it work, they would have to shrink the 550,000-square-yard factory into 7,100 square yards. Theoretically (very theoretically) it was possible.

"We arrived at the very last moment," Baldini said, polishing off a bittersweet chestnut flan with the swipe of his fork. "If we came one year before, they would have had nothing available, because production still would have been going on. If we came a year later, everything would have been sold off or destroyed." They struck a deal with the Ligurian government to occupy the LRF building for free, with the government paying to restore the structure and the necessary mechanical systems. The two would basically transform the LRF into a self-sufficient twenty-first-century FILM Ferrania, capable of housing the entire film-making process under one roof. Only this, Baldini said, could guarantee one hundred more years of analog film.

It was mid-April. The larger factory next to the LRF (which they'd nicknamed "Big Boy") was scheduled to be torn down sometime in June. Here is an abbreviated laundry list of what needed to be done before the first roll of new Ferrania color-reversal slide film could roll off the assembly line: All relevant machinery needed to be purchased, disassembled, and removed from Big Boy before it was demolished, warehoused, reassembled in the LRF, tested, and recalibrated. Anything else had to be located somewhere in the world and shipped to Ferrania. All the necessary chemicals needed to be secured, safely stored, then tested and reformulated. Every important document, from the prized original film formulas and their myriad ingredients to the blueprints of every single machine, widget, system, and building, needed to be located, filed, and backed up. Ferrania Technologies was going to disconnect the steam, electricity, and water powering the LRF, so new self-sufficient systems had to be purchased and installed. Asbestos needed to be removed from the building. The leaking roof had to be patched. Lights needed to be reconnected; twenty-year-old computers, restarted. The toilets required a good scrubbing. There were a million dire tasks, major and minor, and all of it fell on the shoulders of Pagni, Baldini, and a handful of former FILM Ferrania employees they had rehired.

"The risk here is that you get lost," Baldini said, as he and Pagni climbed the dark staircase in the LRF (which had no heat on this damp and chilly afternoon) to face another day of limitless challenges.

One worry that did not keep Baldini and Pagni up at night was whether anyone would eventually buy their film. Based on information pulled together from various news stories, company filings, and industry reports, Baldini estimated the global market for 35 mm still photography film had stabilized at 100 million rolls annually in 2015. Combined with processing services, this still represented a $1 billion global market. "In the future, I believe the market will rise up a bit more, like vinyl records," he said. The more Kodak discontinued beloved products, such as Kodachrome, the more opportunity there would be for Ferrania's film to carve its own niche. Even if they manage to capture only the 4 percent of the global film market that Ferrania held at its peak, it would be a healthy business worth millions annually.

"The market is stable now," Pagni said, "but photographers aren't excited because there are no new products, and only old companies pulling back products. There's no momentum." People don't want to invest in a dying idea, even if they love it, but they will readily pour money into something that seems to be growing, especially if it is against the general trend. In analog photography, that precedent had already been well established by the folks at Lomography.

———— ◆ ————

*I*n 1984, a Soviet military company in Saint Petersburg launched the Lomo Kompakt Automat (LC-A), an affordable plastic camera for the proletariat. The LC-A quickly grew popular across the communist bloc, capturing family memories and vacations from Vladivostok to Havana. Fast-forward to 1991. The Soviet Union has collapsed and Eastern Europe's borders are open. At the edge of all of this was Vienna, a city known for its art and culture, which served as the contact point during the cold war between East and West. Suddenly, Vienna was a portal into the newly lifted Iron Curtain. Within an hour's drive you could be in Slovakia, and in less than three you get to East Germany, the Czech Republic, Bosnia, Croatia, Slovenia, Serbia, Romania, Poland, and Ukraine.

That spring, a small group of students from Vienna, including Sally Bibawy and her boyfriend Matthias Fiegl, took a weekend trip to Prague. In a camera shop they discovered an LC-A, and thought this funny-looking, cheap camera would make a great souvenir. Back in Vienna, they developed the film from their trip and were astounded at how different the photos turned out, compared with those taken on more precise Japanese and German cameras. The Lomo's photographs were saturated with light and color, and had darkened edges and unpredictable variations in tone. The combination of Soviet mass production and plastic components meant that light leaked all around the camera. The LC-A took photos that were decidedly imperfect, but Bibawy, Fiegl, and their friends fell in love with its aesthetic.

Soon, they were spreading the gospel about the LC-A, buying as many as they could, taking the cameras everywhere, chewing through rolls of film. Instead of picking out the clearest, best shots, then enlarging and framing them, as most photographers did, they just printed off whole rolls at a time at the cheapest supermarket kiosks, and pasted every picture up on the walls of their apartments, even the blurry ones. It was experimental, somewhat surrealist, and radically different than what photography had become.

"The early years of Lomography was about the transition of a visual understanding," said Bibawy as we sat down to lunch, with Fiegl and their two sons, at a restaurant in the hills above Vienna. "Photography was broken. It was all so rigid and precise. There were too many rules. We said, 'Forget about quality.'"

In the summer of 1992, Bibawy, Fiegl, and friends formed the Lomographic Society International, and published a manifesto in a local newspaper, which featured ten rules of Lomography, their new philosophy of photography that embraced imperfection in all its glory. "Lomography is a fast, immediate, and unashamed form of artistic expression," the manifesto read, which instructed others to (1) take their camera everywhere, (2) use it anytime, (3) shoot from random angles, (4) shoot up close, (5) shoot without thinking, (6) shoot quickly, (7) embrace the unpredictability of shots before, (8) and after, (9) embrace the camera as part of life, and (10) always ignore the rules. Lomography was a liberation philosophy for photography, rather than a technical discussion aimed at achieving the perfect picture.

As demand for the LC-A camera increased, Lomography quickly grew into a business for Fiegl and Bibawy. By 1995 they were the LC-A's official international distributors, and Lomography launched a website that invited users to scan and upload their photos, creating one of the first online photo-sharing platforms. Over the following years they opened various stores around the world and introduced new cameras, including the Actionsampler (which took four simultaneous photos on four different lenses); the nine-lensed Pop9; the Colorsplash, with different color filters and flashes; and dozens more models. From 1995 to 2001, Lomography's business grew, on average, 50 percent a year.

Although the digital photography revolution was well under way at that point, Lomography was somewhat immune to the decimation of the film business. Its cameras were selling, and the company kept opening more stores, on every continent. But in 2005, the original Russian LC-A factory announced that it was ceasing production. Bibawy and Fiegl decided to make a new version in China, but they faced a choice. Would the new LC-A, the flagship of Lomography, still be an analog film camera, or should they make a digital version? They put the question to the Lomography community online, and in a survey asked it to assign emotions and experiences associated with *digital* and *analog*. The results intrigued them.

"All the emotional words were put into the analog field," Fiegl said, "while the digital field was all about perfection and speed." The survey results were conclusive. Ninety-five percent of Lomographers wanted an all-analog LC-A. This not only shaped the new camera, but the company's direction. Lomography went all-out on analog. Lomography quickly adopted the slogan "Film's Not Dead," and the next year at Photokina, the global imaging industry trade fair, they flew a huge banner proclaiming "The Analogue Counterrevolution."

That really lit a fire. From 2005 to 2010, Lomography grew even more dramatically, releasing more cameras each year in every possible format (medium-format cameras, instant cameras, cameras that spin around, cameras with crazily colored cases . . . more than 150 different models) and expanding their retail presence. Today, the company does roughly €20 million in sales each year, and though it has closed a number of stores in different cities following an overly aggressive retail expansion, there are still more than forty franchise and corporate Lomography locations worldwide, and the company remains profitable. "Our business has been very stable in the long run," Fiegl said.

Rather than weaken it, the rapid growth of digital photography seems to have fed Lomography's success. It has driven younger photographers to Lomography products (the average customer is between 20 and 45 years old), and has allowed them to charge a premium for cameras that are, in the words of one photographer I know, "crappy plastic toys" that cost as much or more than a much higher-quality secondhand Canon or Nikon. This is especially poignant when you consider the company's undeniable influence on digital photography's look and feel. From the blurred images to the saturated filters, randomly placed shots, and even the practice of sharing your images online in social networks, the vernacular of modern smartphone photography (especially Instagram) is almost a textbook adoption of Lomography's ten principles.

⁎

*B*aldini and Pagni see Lomography as proof that the analog film market will respond to new ideas and products. Lomography's customer is not a nostalgic one, holding on to film out of fear or stubbornness, but a photographer looking for new experiences that differ from the digital standard without getting too technical. "If you browse an online newsgroup on analog photography, it is lots of fifty-year-old men talking about dimensions of grains and blah blah blah . . . it's fucking boring!" Baldini said in disgust. "Lomography gets the twenty-year-old who never heard of an analog camera and buys one after hearing the shutter click for the first time."

To Baldini, as a photographer and filmmaker, it is not a question of film versus digital, or a desire to achieve a better resolution, or clarity, or other measurable image quality with one or the other. He has a huge collection of digital cameras, and considers his iPhone the best possible camera with which to take photos every day. Film was a choice, and those who chose to use it (a population that still numbered millions of professional and amateur photographers worldwide) did so because they loved something about the analog process and the look it produced. Film photographers wanted a more hands-on relationship with their material.

"It's just like art," said Baldini. "I'm Michelangelo and I have the idea to do a sculpture of David. I have two choices today: I can scan David's body and print out the perfect proportions on a 3-D printer, or I can start from a block of marble and chip away. The process is different and I have to be more creative with marble to achieve a result that becomes a masterpiece," he said. Film would continue as a viable creative tool, just as paint and canvas did long after photography came along. "When we sell a film, I want to sell a supportive tool to obtain a creation."

The way to do that was to set up FILM Ferrania as an "artisanal/industrial" operation, creating professional, high-quality film products that have their own distinctive look and feel. In the heyday of the film business, every brand had this. Kodak was known for its reds and oranges, Agfa for its greens and blues, and Fuji for a neutral balance. Film needed a character, a sense of place, a terroir like the slightly fizzy red wine made nearby, which changed and evolved over time. "It can't be good for everybody," Baldini said. "For everybody, there's the iPhone."

We were standing by the entrance of the building they called Big Boy, the original FILM Ferrania coating facility that Baldini and Pagni were stripping for equipment before its demolition. Big Boy was an imposing five-story windowless concrete bunker the length of a city block. We waited in the bare lobby for our escort, Paolo, a former employee now doing site maintenance, who showed up in a white cleaning suit, bearing three flashlights. We climbed the staircase and passed offices that had been frozen in time. There were calendars on the wall and yellowed family photos still taped to monitors. Papers spilled off desks, keyboards sat under 2 inches of dust, and peeling paint piled up on the floor like dirty snow.

We went deeper into the building, until we entered darkness. The entire factory, in effect, is a giant darkroom. "The building is the machine," Pagni said, as we followed a faint strip of luminescent paint on the ground. "You cannot separate them." The experience of walking through a massive, perilous, abandoned factory in the dark, lit only by flashlights, is fascinating and terrifying. "Just imagine working here," Pagni said, noting that many of the former Ferrania employees he met told him their personalities radically changed the day they entered eternal night.

"You are likely the last outside person to see this place set up for production before it is torn down," he said. What I saw was very limited, but it gave me a basic understanding of the incredibly complex way that film is made. First, a thin sheet of transparent base material is created out of cotton cellulose fiber and other chemicals, and rolled onto huge spindles called Jumbos. Next, the specific formula that produces the look, feel, and performance of the film, called an emulsion, is mixed. Each emulsion is made from a gelatin base (Kodak once owned a cattle ranch in Colorado, to provide a consistent supply of bones and hooves for its gelatin) and a complicated cocktail of different chemicals, silver particles, and other substances.

The transparent base is unspooled from the Jumbo into a coating machine, which layers emulsion onto the base in a consistent spray, with each color locked into place by electrolysis. Every emulsion is sensitive to a different color, so you need multiple emulsions for each type of film. Black-and-white film might need three emulsions, whereas color film could use up to sixteen. After the emulsion is applied, the sheet of film passes through a blast freezer, and then a drying tunnel. This is a series of ducts that carry the film forward on a steady cushion of air, looping up and down more than a dozen two-story tunnels at nearly 2 miles an hour. All of this has to be done in one continuous motion, with the film unspooling and running through the process without stopping, like a movie rolling through a projector, but on the scale of a building. The coated film is rolled back up, "seasoned" over a period of time (this determines its speed), and taken to converting, where it is cut into the correct size for its format by the perforator. Finally, the film is fed into another machine, which rolls it into canisters, and then drops each roll into a plastic case and cardboard box.

I reached into the machine and pulled out two rolls of Lomography film, which were the last products being produced here when the plant shut down four years earlier. Every machine in the building still worked, and had cost untold millions to design, manufacture, and install, but they were completely worthless without the custom software, manpower, or supporting suppliers associated with them. Whatever Pagni and Baldini didn't rescue and take next door would be discarded.

Back in the LRF building, Pagni demonstrated how they hoped the new operation would work and the challenges in making that happen. Thanks to 3M's original design, the LRF had a built-in coating machine and drying tunnel, which could produce up to 8 million rolls of film a year, compared to the machines in Big Boy, which were built to produce hundreds of millions of rolls. "Conceptually it's all very easy," Pagni said, after explaining each component of the LRF coating machine, "but behind it is *chemistry*, and that's where things get complicated."

Pagni and I continued our tour throughout the LRF building. The computer room, which controlled the various precision systems, from heating and cooling to the chemical mixers, sensors, and the coating machine itself, was a return to Radio Shack circa 1991. The place ran on vintage IBMs, HPs, and all manner of beige DOS relics, most with floppy disk drives, and few with anything more recent than Windows 95. "The formula for automated manufacturing is stored inside these computers," Pagni said, patting a desktop that was emitting a steady crunch. Next, we sifted through drawers of blueprints, formula binders, and stacks of microfilm, which contained all the pertinent information for the emulsions, machines, and the building itself. "If you lose these," Pagni said, "you have nothing." The knowledge of FILM Ferrania was scattered, buried, and hidden in those fragile documents.

The problem was, they couldn't just start fresh with modern computers and equipment. They were trying to wake up a sleeping giant, and that giant had run on this old gear, and run well. Any sudden change would add another variable to a process brimming with them. Everything required custom-made, highly specialized sensors and instruments specifically calibrated for this building. "Look at this," Pagni said, walking over to a machine that looked like a giant milk shake mixer. "This determined the amount of silver particles in a particular emulsion by spinning at a particular speed. The formula that we have is calculated for this one specific machine here. You can't easily reverse-engineer that with a new machine!"

The potential for failure was massive and held serious consequences beyond their investment. "There is no margin for failure here," Pagni said. "If you fail at something in this process, you can hurt people. If I turn off the blower by accident when my chemical engineer is working with ammonia, he'll be dead." He paused for a second, rubbing his beard, before looking up. "It's heavy, the intellectual weight you need to carry to make this."

———•———

Few people could appreciate that weight, but Florian "Doc" Kaps was one of them. "We were in the same situation as Ferrania," Kaps told me over dinner in Vienna the following night. "We know the shit storms that are coming up. Should I tell them to run, or go do the impossible?" Kaps, who wears his hair in a ponytail and always has a mischievous grin, is best known as the founder of the Impossible Project, a company that began producing film for Polaroid cameras after Polaroid discontinued its own film production. A trained biologist who studied spiders' vision, Kaps joined Lomography in 1999 and worked with them for several years, but when Kaps felt strongly that the company should pursue instant photography, he struck out on his own.

At its peak, Polaroid was the equivalent of Apple: a golden technology star whose every innovation was more astounding than the last. But by 2004, Polaroid was emerging from bankruptcy, mired in mismanagement, and had squandered its cultural capital. Digital, of course, hadn't helped things, and Polaroid film production was on the ropes. Kaps wanted to apply the same analog social marketing he had experienced at Lomography to Polaroid film, but Polaroid couldn't care less. He ordered more than $200,000 worth of film, and set up an online store called Unsaleable. "Polaroid did their best marketing for us," Kaps said. "They closed down, year by year, and started getting rid of all formats, which I bought the last of." Polaroid had planned for a continued decline of the film market over the preceding decade, but despite this, demand kept up, at around 25 million packs of film a year. "Everybody thought digital would kill analog, but suddenly people started missing the touch of film," Kaps said. "They initially thought the biggest problem of digital photography was quality, and as soon as quality improves, it'd win. But the biggest problem of digital photography is that it isn't real. Photographs disappear, and the amount of pictures that turn real is decreasing dramatically. There's no family albums anymore, no prints anymore, nothing you can touch or shake. And people started to miss that."

I could easily relate. Since I switched to digital photography, I have taken more than fifteen thousand digital photos, but printed just a tiny fraction of those—maybe half a dozen albums and a handful of pictures. Digital photography actively discourages the physical artifact of a photograph. Taking the photo is simple with digital cameras, but everything after that is a tremendous effort: editing, formatting, sifting, price-shopping, layouts, more formatting, more editing. It took my wife and me nearly two years to print our wedding album. Instant film photography delivers the best of both worlds: digital's instant gratification with analog's physical artifact.

Then, in 2008, Kaps received a distressing message from Polaroid. The company invited him to attend the June 14 closing of its last operating film factory in the world, in Enschede, the Netherlands. Kaps freaked out. Unsaleable was making hundreds of thousands of euros each year, shipping rare, even expired, Polaroid film around the world at more than twice its original price. If the Enschede plant closed, that would be the end of Polaroid film. Two hundred million Polaroid cameras would become paperweights overnight, and Kaps's business would quickly dry up.

All summer long, Kaps harangued Polaroid to let him buy the factory, but the company was deaf to his offer. Then, on September 24, the FBI and IRS agents raided the offices of Petters Group Worldwide, the Minnesota investment company that had owned Polaroid since 2005 but was actually running a $3.4 billion Ponzi scheme. Kaps got a call from someone who had been granted power of attorney over Polaroid's affairs. "You have one week," he told Kaps. Kaps scraped together €180,000 to buy the equipment in the Dutch plant, and took over the building's lease. He bought all the remaining Polaroid film stock, and used the profits from its resale to finance the revival of the plant, at a total cost of €4 million. The new company was named after a quote from Polaroid's founder, Edwin Land: "Don't undertake a project unless it is manifestly important and nearly impossible." The Impossible Project was born.

Here is where Kaps's fairy tale ended and the reality of reverse-engineering Polaroid film took over. "We started to find out everything was more difficult than we thought," he said, laughing at his own naïveté. Instant film happens to be one of the most complex products in imaging. If the 35 mm film from Ferrania is a sturdy, timeless Fiat, then Polaroid is a precision-tuned Ferrari. Similar in theory, but incredibly difficult to pull off. Essentially, each Polaroid is a sandwich of twenty-six different layers, which are exposed and squeezed together as they exit the camera, triggering the photograph's chemical development. The Impossible Project factory in Enschede was responsible for assembling these twenty-six different components from across Polaroid's manufacturing network into packs of film. The problem was that each of those twenty-six ingredients came from different factories and companies spread all over the world. Many of these had closed down or discontinued production. Polaroid was a vast company that spent billions of dollars in research, employed thousands of engineers, and ran on complex systems of quality control and procurement. The Impossible Project was five guys in a factory who only understood the last part of this long, complex process.

The first Impossible Project film hit the market in the spring of 2010. "Ugh," Kaps said with a shrug. "Let's just call it 'experimental.' We delivered film that was worse than any nightmare we'd ever had." It barely worked. Photographs were blurry, splotchy, and filled with weird shapes, if they captured anything at all. It was so sensitive that the second you took a photo, you had to snatch it from the camera and put it in a dark place, and keep it warm next to your body (though not against skin, because the chemicals sometimes leaked). "It was a disaster in the US," Kaps said. "Americans were used to Polaroid film at nine dollars a pack, in bright colors, that behaved." Polaroid film took a minute to develop. The new Impossible Project film took nearly an hour, if it worked at all. For this crapshoot, customers paid $25 for a pack of eight pictures.

Instead of delivering a new Ferrari, the Impossible Project was selling an old Alfa-Romeo, a beautiful dream of a thing riddled with problems. But a lot of people still buy those old Alfas (Baldini drove one), and a lot of people bought the Impossible Project's imperfect film. This is the key lesson Baldini and Pagni have drawn from the Impossible Project, and one that's central to the success of new analog brands facing digital competition. The Impossible Project succeeded precisely *because* it focused on celebrating analog film's imperfection, rather than chasing digital perfection. Kaps turned the quirks into selling points. New analog photographers, especially younger ones trying out instant photos for the first time, embraced the randomness and unpredictability of the film in the same way Lomographers did with their cheap Soviet cameras. We can easily have seamless reproductions with digital, but that turns out not to be what we all want all the time. They bought these cameras for the promise of one-of-a-kind photographs, like the slightly blurry portrait of Taylor Swift's torso (cut off at her eyes) that graces the album cover of *1989*, which was taken with Impossible Project film. The Impossible Project created a messy, defective product that succeeded because it was the furthest thing from perfect, the antithesis of digital photography.

The film's performance has slowly improved, and the Impossible Project now sells over a million film packs a year (each pack still yields 8 photos), with sales growing around 20 percent annually. The company also buys up all the old Polaroid cameras it can get its hands on, refurbishing and reselling them for $200 to $400 each. In 2012, Kaps sold the controlling interest of the company to Slava Smolokowski, a self-made Polish commodities billionaire, who was introduced to the Impossible Project by his son, Oskar. Kaps soon brought Oskar Smolokowski onboard as his assistant, and he joined us for dinner in Vienna in April 2014, shortly after assuming the title of Impossible Project's CEO at just twenty-five years old. This past spring, Smolokowski launched the Impossible Project's first new camera, which features Bluetooth-enabled controls that you can operate from a smartphone (mixing analog capture and digital flexibility), five lenses, and an adaptive flash. Still, Smolokowski estimates, the Impossible Project owns just a fraction of the instant film business that Fujifilm dominates with Instax.

Instax launched in Japan in 1998 as a smaller, less expensive alternative to Polaroid (made using technology Fujifilm had licensed from Polaroid), and the product was aimed mostly at Japanese schoolgirls, complete with accompanying stickers. Sales peaked in 2002 at a million cameras, according to Yoshitaka Nakamura, the head of the Instax group, then plummeted to 100,000 cameras by 2005. Fujifilm drastically downsized production and nearly killed Instax off. Instax sales turned around starting in 2007, and Instax has grown dramatically since then, from 1.6 million cameras sold in 2012 to more than 4.6 million in 2014 (the year a Hello Kitty version was released), and likely more than 6 million in 2015. Instax is now the world's most successful photographic film, selling what Smolokowski estimates to be upwards of 40 million packs, a number he says is closing in on half the film Polaroid sold before it shut down film production in 2008. At a time when Fujifilm's digital camera sales have been decimated by smartphone cameras (in 2008, Japanese manufacturers shipped 110 million digital cameras; in 2014, they shipped just 29 million), the Instax unit was credited with pulling Fujifilm's imaging department back to profitability in 2014. That's right, film (and Hello Kitty) saved Fujifilm from its own digital slump.

More new ventures are filling this niche instant film market, including New 55, another Polaroid revival product for a specific type of discontinued medium-format Polaroid film that hit the market last year. The instant film revival has even spurned the development of new digitally enabled products, such as Polaroid's own point-click-print Snap camera, and Prynt, a photo printer that attaches to a smartphone and uses digital thermal printing technology (originally developed by Polaroid) to produce images. Prynt isn't strictly analog, but it does provide a novel physical solution for the most used cameras. Clément Perrot, the company's cofounder, came up with the idea at a party in San Francisco, when one of his friends was shooting pictures with Impossible Project film. "We felt that among our generation, you end up taking hundreds of thousands of pictures," Perrot, who is twenty-five, told me. "But the more pictures you take, the more meaningless they become. You post them online and they end up being deleted by a whole feed of digital information."

Other legacy film companies have seen new life postbankruptcy. British black-and-white specialist Ilford was purchased in 2005 by its former managers and reborn as Harman-Ilford, with production resized to suit the current niche market. They developed new products and quickly turned the company around. The company's sales in 2014 were reported to be £30 million, and it has been consistently profitable since restructuring. ADOX, based in Germany, does a similar scaled-down production, including film for Lomography. It boasts that it has built the smallest photochemical factory in the world. "The heydays of analogue photography are over," its website says. "Our goal is to manufacture the finest quality analog product with the lowest possible investment in equipment and buildings so that we stay flexible and small enough to live off future market volumes."

Even Kodak, which produces about 1 percent of the film it once did, seems to have found a renewed interest in analog, especially in motion picture film (the part of the business that US-based Eastman Kodak retained after bankruptcy). Fujifilm exited the cinema business in 2013, leaving Eastman Kodak as the only major provider of movie film in the world. That was a tenuous position. Most movies and TV shows are now shot and projected digitally, and the death of Kodak's cinema division looked inevitable until Hollywood came to the rescue in 2014. A group of the industry's biggest directors, led by J.J. Abrams, Quentin Tarantino, Christopher Nolan, Judd Apatow, and Martin Scorsese, aggressively lobbied the studios to purchase enough Kodak motion picture film to guarantee a continual supply. The attention this generated seemed to invigorate Kodak's commitment to film, which it publicized relentlessly online and in the press. This past January, Kodak announced they would begin making a new version of their iconic Super 8 film camera later this year.

"It's not just an idea that would be nice to keep going," Abrams, the director behind such blockbusters as the film-centric *Super 8*, *Star Trek*, and *Star Wars: Episode VII—The Force Awakens*, told the *Hollywood Reporter*. "It's an aesthetically and materially important thing."

Nolan (*The Dark Knight, Inception, Interstellar*) was more direct. "Film has tremendous balls. That's just all there is to it."

I spoke with Abrams by phone in late 2014, as he was filming *Star Wars: The Force Awakens* in London. "Analog, whether it's film versus digital, or tape versus Pro Tools, or paint versus Photoshop . . . each tool is appropriate for a different task and requirement," he said. Abrams has a fondness for analog tools and ideas. He writes his first drafts in a paper notebook, and cocreated a best-selling novel called *S.* that was an elaborate package of paper, notes, and physical ephemera specifically designed to stand out as an analog artifact, more than just a story in text. When he approached the latest *Star Wars*, Abrams consciously went in the opposite direction that George Lucas had during the second installment of the series a decade prior. Those movies came out when digital effects and filmmaking were cresting a wave of technological superiority, and Lucas reveled in it. But ultimately this digital infatuation detracted from those movies. The characters and sets looked superficial, scarcely distinct from video games, and let's not even talk about Jar Jar Binks.

"We are shooting on film, which is a big deal," Abrams said, noting that when given the choice, he prefers film for its visual texture, warmth, and quality. "But I think that the decision was more representative of the approach [to the movie] than anything. We wanted to embrace an analog approach. That doesn't mean we're not using digital effects, or Industrial Light & Magic won't, but we're also using things not as frequently used today."

The set and props for *Star Wars: The Force Awakens* were largely built by human hands and painstakingly painted, so they wouldn't have the "inherent perfection of the computer generated model." Abrams used more masks, makeup, and physical robots to bring creatures to life, just as Lucas had back in the 1970s. This was important, because it shaped how the film would look to audiences, as well as the actors' performances. "I've done a number of movies with creatures and things that aren't there," he said. "I can't explain it to you other than to say I can tell when someone is performing something in physical makeup. Even the crew begin treating this creature differently because it's *there* . . . rather than someone trying to make believe there's something there."

———————◆———————

Abrams's *name came up when* Baldini spoke about FILM Ferrania's campaign on Kickstarter in late 2014, which had collectively raised $315,000 to make the company's goal of securing "100 More Years of Analog Film" a reality (apparently, Abrams was among the project's 6,000 backers). That money had been used to finance the purchase of the equipment from Big Boy and other buildings, and to begin preparations for the move to the LRF. So far, Baldini and Pagni, with additional support from some friends and family, had already invested €1 million in the project, and they were actively looking for another million from outside investors, to be able to produce around 300,000 rolls of film in 2016. (In fact, Ferrania's first new batch of film, P30 black-and-white 35mm, was shipped to early backers in February 2017).

They had big dreams for FILM Ferrania: cameras, a world-class film research lab, an archive for restoring Italian cinema gems, and automated film-processing services that put developing a roll of Ferrania film just a swipe of the iPhone away . . . but for now, they were sitting in the near dark in the heatless LRF with their chief chemist, Corrado Balestra. Every day they were putting out a million little fires while they waited for the government to remove the asbestos in time for the demolition of the larger buildings, all before the power and water were cut off and their chemical emulsions from the first test batch expired. Already they had sent updates to their Kickstarter backers about the delays, and the responses were universally encouraging and supportive. "Don't give up!" wrote one. "Let me know if you need help, I'll come over!" commented another, though as FILM Ferrania's delays mounted over months, that patience frayed. They were deep, deep into it now, and still a year or more away from that first elusive roll of film, but no one expected FILM Ferrania to run as a normal business or produce a perfect product.

"No one thinks two young guys with some small experience and some small capital have taken over one of the biggest industries in the region," Pagni said, leaning back in a creaking desk chair that was quite literally on its last legs.

I asked them what it would feel like to take that first roll of film off the line, put it in a camera, and snap a photo.

"Revenge!" Balestra said, with a wicked chuckle.

"No, the first picture will only be the start of our real challenge," Baldini said soberly. "I don't want only one."

4

The Revenge of
Board Games

"*H*ey, Ben," *said Aaron Zack,* looking up from his laptop screen at Ben Castanie, his business partner, "Looks like another board game café is opening downtown."

I had just sat down to interview Castanie, owner of the pioneering Toronto board game café Snakes & Lattes, when Zack, the company's director of operations and co-owner, shared this news. Castanie, a thirty-four-year-old thickly built Parisian with a whisper of a beard, graying dreadlocks, and wrists jumbled with bracelets, just shook his head with a bemused smile and shrugged. He had gotten used to hearing this with increasing regularity.

It was a bitter Tuesday morning in March, a month before my visit to Italy, but Snakes & Lattes was warm and bustling. The espresso machine hissed, and laughter rose up from a dozen tables. By lunchtime, the café, which seats around 120, would steadily fill up, and by six that night, every table would be occupied. At that point, the bustle in here would transform into a deafening tumult: a mix of belly laughs, defeated groans, surprised screams, triumphant shouts, and the click-clack of plastic and wood on cardboard . . . all set to the soundtrack of classic pop music, which wouldn't peter out until well past midnight.

On the weekend, the crowd at Snakes & Lattes spills onto the sidewalk, and the wait for a table can stretch up to three hours.

Snakes & Lattes had to develop its own reservation software just to cope. It has already expanded the space to more than double in size (it originally seated closer to 50), and opened a second location just over a mile south, in a former pool hall that is double the size of the original café with room for hundreds more customers. The tables are wobbly, the chairs uncomfortable; there is no Wi-Fi, and the food . . . well, let's just say it's not a draw. For all this, you pay $5 a person just to get inside.

What brings hundreds of thousands of people to Snakes & Lattes each year are the best board and card games in the world, from such classics as Jenga and Scrabble to modern hits, including Set-tlers of Catan, and brand-new independent games, such as But Wait, There's More! Snakes & Lattes is a mecca of analog fun, an all-you-can-play smorgasbord of gaming, and an example of how a tangible community is closely tied to analog's revenge.

Unlike paper correspondence, vinyl records, and photographic film, board games and card games (both tabletop games, in industry terms) were not overturned by digital competition. While there has been an overall decline in the tabletop game industry from its heights in the early 1980s, due largely to the growth of video games, it is not as though the tabletop industry nearly vanished or its major players all went bankrupt. I grew up during the decades of tabletop's supposed decline, and I fondly remember marathon Monopoly tournaments, summers filled with games of Cranium, and endless lunch breaks in fifth grade playing Balderdash. I cannot recall a single TV commercial for a Nintendo game, but the jingles for Mousetrap, Guess Who? and Hungry Hungry Hippos are seared into my memory.

What happened to tabletop games was more of a steady slide, affecting sales, but more important, quality. Aside from a few once-a-decade hits, such as Cranium, board games became a stale category. The larger companies, such as Hasbro and Milton Bradley, preferred to focus on new editions of blockbuster titles (Monopoly: The Bea-tles Edition, Monopoly: Angry Birds, Monopoly: Electronic Banking, and so on). The innovation was found in video games: first in arcades, then on personal computers, consoles, and mobile devices. As Inter-net bandwidth increased, a world of real-time, globally connected, massive multiplayer gaming opened up endless possibilities. When you could command a squadron of tanks into Nazi Berlin from your living room, or play *Words With Friends* on the bus against an opponent in another country, how could Risk and Scrabble possibly compete?

This has indeed been the pattern in the gaming industry. Toy store shelves groan under the weight of Monopoly, Risk, and Scrabble spin-offs, and their sales remain anemic. But over the past few years, something dramatic has transpired in the independent, player-driven corner of the tabletop gaming industry referred to as the hobby segment. Hobby game sales in North America alone have more than doubled since 2008, posting double-digit growth every single year. Once the awkward stepchild of the North American tabletop game industry, today the hobby game market makes up nearly half of the game and puzzle segment's $2 billion revenue for the toy industry, a huge shift from just a few years ago. The Game Manufacturers Association (GAMA), a US industry group, has more than doubled its membership of companies since 2009. Attendance at gaming conventions, such as GAMA's Origins, GenCon, and Germany's Essen Spiel, is consistently breaking records for both professionals and gamers.

There are countless new tabletop games, as well as publishers, designers, blogs and podcasts, stores, and yes, game cafés, opening up to serve this growing market. News articles appear with regularity proclaiming that "Board Game Sales Soar" (NPR), "Board Games Are Growing in Popularity Again" (*LA Times*), "Board Games Are Back" (*Fortune*), and that we are living in a "Board Game Renaissance" (*The Guardian*). Tabletop games are booming, in both sales and cultural relevance, and it is happening precisely because of the inherently analog nature of board games and the unique social need they fulfill in our lives.

Games were the bright spot of Ben Castanie's childhood. He grew up in a poor housing estate in Paris's outskirts, and though his family was far from destitute, his parents did not spend money on games. "Our parents were allergic to play," he said. That changed each summer, when Castanie and his siblings stayed with relatives in Marseille and discovered the free toy lending library, called a *ludothèque*, nearby. These public institutions were a dream for kids, stocked with everything from Barbie dolls to LEGO, and vast shelves of games. Castanie spent his summers immersed in Scotland Yard and Monopoly, and endless card games, such as Mille Bornes.

In 2008, Castanie was working for a logistics software company in Toronto. On a weekend trip to Chicago with his girlfriend at the time (a woman he'd known since childhood), Castanie happened into a game and toy store. The couple didn't buy anything, but the brief visit sparked a conversation about the wonderful ludothèques of their youth. "We thought 'This is totally missing in North America,'" Castanie recalled. "'Why don't we create a toy library in Toronto?'" Quickly, they realized that a ludothèque meant dealing with hundreds of kids daily. Maybe a board game café would be more suitable for two people in their twenties.

Castanie had no experience with board game cafés, but he began researching online and saw that there had been a rapid explosion of them a few years before in Seoul, South Korea, beginning around 2002, and fading away just a few years later. Germany, where board games were very popular, had many bars and cafés with games, and there was even one in São Paulo, Brazil. But when he researched North America, the only thing that seemed to exist in large cities were hobby and game stores. Castanie despised these stores. They were usually in basements or on second floors, packed with miniature fantasy figurines, staffed by grumpy middle-aged men whose only setting was sarcastic disdain, and attended by hard-core gamers who came to roll the ten-sided die of Dungeons and Dragons or shell out their precious dollars for packs of Magic: The Gathering cards.

These stores had sucked the life out of board games in North America, with their elves and wizards, their desperate male loneliness, and their sneers of nerdy exclusivity. Castanie was determined to win it back. "I'm against hard-core gamers and geek culture," he said. "My number one goal was to welcome everyone, and make this place as far away from geek culture as possible." Castanie would build a place that married the selection of a European game library with the accessible vibe of a neighborhood coffee shop. Good coffee and games and nice people. Board games for the mainstream, for the everyman and (crucially) everywoman. It was a remarkably simple formula.

Over the next two years, Castanie and his girlfriend collected games. He would stop at the Salvation Army and other thrift stores after work, and scoop up entire shelves of games. He drove to nearby towns and cities, answering Craigslist ads from people dumping their board game collections. Over two years, Castanie and his girlfriend (who is no longer involved with him or the business) acquired a thousand games for less than $2,000. With $50,000 borrowed from their mortgage, the two secured the lease on a grimy, cigarette-stained doughnut shop in Toronto's Koreatown neighborhood, right by the university student retail strip called the Annex.

Snakes & Lattes opened on a Monday just before Labor Day 2010. When Castanie peeled the construction paper off the window at 11:00 a.m., there was already a line outside. The hard-core gamers he so desperately wanted to avoid were his first and most eager customers, but the crowd quickly diversified, and Snakes & Lattes was an instant success. I remember it pretty clearly, because it opened just two blocks from the apartment I once shared with my friend Adam Caplan. Every time I walked by the place, it was absolutely packed. Certain subcultures quickly took to Snakes & Lattes as their preferred hangout. It didn't serve alcohol, which attracted young people looking for a fun, wholesome alternative to bars and clubs. These included large groups of Asian students, orthodox Jews who were going out on group dates, Muslim teenagers and couples, groups of young women, and countless dating encounters, who filled up half the tables.

"The community is less centered on board games than on a cool hangout, a place to have fun," Castanie said. "People don't come here necessarily to play board games, discover them, or even buy them." He estimated that less than 10 percent of customers at Snakes & Lattes were hard-core gamers. That group primarily plays at home or in hobby stores. "When we concentrate too much on board games, we lose a little bit of our focus." Tabletop games are just an excuse for getting together, but a perfectly designed, uniquely suited one, specifically because of their analog nature.

Tabletop gaming creates a unique social space apart from the digital world. It is the antithesis of the glossy, streaming waterfalls of information and marketing that masquerade as relationships on social networks. A Twitter conversation is nothing more than a chain reaction of highly edited quips; a Facebook friendship is more like an electronic Christmas card exchange than a real interaction; an Instagram feed captures just the shiny highlights of life. "Networked, we are together, but so lessened are our expectations of each other that we can feel utterly alone," wrote MIT professor of sociology and psychology Sherry Turkle in her book *Alone Together.* "And there is the risk that we come to see others as objects to be accessed—and only for the parts we find useful, comforting or amusing. Once we remove ourselves from the flow of physical, messy, untidy life. . . . We become less willing to get out there and take a chance."

Playing a board game in a neutral environment, such as Snakes & Lattes, transforms the way its players relate to one another. They engage. They speak. They laugh. They embrace vulnerability. They are human. Snakes & Lattes quickly became what sociologists call the third place for many of these groups: a safe, welcoming, sacred space outside of home and work, where people are free to explore the bounds of identity and human relationships. "A sacred space is not a place to hide out," Turkle wrote. "It is a place where we recognize ourselves and our commitments."

Video games were supposed to deliver these interactions, and at one point they did. I have extremely fond memories of sitting next to my brother Daniel in our basement, jumping and screaming as we played *Super Mario Bros.* and *NBA Jam*, daily *Street Fighter 2* battles with my friends Josh and Dan during eighth grade lunch breaks, and an entire semester at university spent throwing knives and spraying bullets at my roommates with *GoldenEye 007*. Whether we played video games in grimy arcades or in our own homes, these became highly social experiences . . . physical, visceral, and certainly bonding.

But as technology improved, video gaming became a solitary experience. Even if you were playing *World of Warcraft* or *Call of Duty* with the same group of friends around the world each day, talking smack over your headsets, and typing in snippets of conversations, you were ultimately alone in a room with a screen, and the loneliness washed over you like a wave when the game ended. By the time the iPad came around and truly mobile gaming blossomed, the last crumbs of real social interaction disappeared from video games. Now, you could lie in bed next to your lover and completely ignore their presence, perhaps even while making virtual love to your virtual lover in *Second Life*.

This very human need for social interaction lies at the heart of the revenge of tabletop games. "Individually, then collectively, we realized the virtual world could never provide us with enough bandwidth to associate with each other the way we want," said Bernie De Koven, a pioneering computer game designer, theorist, and writer who focuses on the study of play and fun. When we play with a computer, either alone or in a multiplayer game, we share ownership of that experience with the software. The program and device restrict our ability to shape the experience of play to our imagination, even in games as pliable as *Minecraft*. "There's never going to be a virtual environment as completely engaging as the physical environment is. It is so much more engaging to play a game of chess face to face than it is online. Online is a good substitute when we can't meet. The ultimate contest is when you see each other face to face; see each other sweat and squirm."

With analog gaming, whether it is an intricate board game or a child's game of tag, all the players need to work together to create the illusion of the game. It requires a collective investment of your imagination in an alternate reality to believe that you actually own Park Avenue, and the colored slips of paper in your hands are worth something. When that happens in gameplay, it triggers what De Koven calls coliberation. "When we're together in a social space environment, we each kind of free the other to be more complete, more full, more yourself," De Koven said. In effect, we work together to liberate one another from reality. That is what the hundreds of people who go to Snakes & Lattes each night are doing. You can do that in the digital world, but only to a degree. "It makes us more alive," De Koven said. "You can't get that from a computer. You can get engaged, you can get excited, but not more alive."

Games can bond us to deep, long-lasting friendships. My grand-parents played bridge with the same three couples each week, my mother-in-law has been clicking tiles with the same gossipy mah-jongg ladies for many years, and my wife now plays mah-jongg with her own friends every month, on evenings that are the highlight of her so-cial calendar (regardless of how often they forget the rules). Whether your game is bowling, softball, cards, dominos, chess, or Dungeons and Dragons, the goal is relationships as much as it is victory.

Evan Torner, an academic who coedits the journal *Analog Game Studies*, believes that a game acts as an alibi for its players to engage in certain behaviors. "I can't invite five friends over to my house and say, 'Let's all play starship!'" he said, aping the tone of an imaginative kid. "But I can invite them over to play a game my friend designed on one card, called Vast and Starlit. It's just this little piece of cardboard that lets us all pretend we're on a starship together easily. Analog parts become the gateway function to a new social world: we adopt char-acters, engage with different economies, colonize a place that appears uninhabited, and have to engage in trade relations. It comes much more simply down to the ability to have social situations from which we're familiar. Analog games do let us work through fiction, but also the [real-world] strategies and tactics underlying fiction."

When people go to Snakes & Lattes, they are seeking what Scott Nicholson calls a rich, multimedia, 3-D interaction with real people. Nicholson is one of North America's foremost game study academics, who now heads up the game design program at Laurier University, outside Toronto. Having extensively studied (and played) both digital and analog games, Nicholson has witnessed a growing desire for real, analog experiences that enhance what gamers do virtually. The *Call of Duty* player will also play paintball, the online poker player travels annually to Las Vegas, fantasy fans will pay to try to break out of an escape room, the *World of Warcraft* regular will don a cape and a foam sword and attend a Live Action Role Play (LARP) retreat.

Analog gaming happens on multiple cognitive levels, which leads to a richer experience. That is obvious when you are running around a for-est trying not to get hit by a paintball, but Nicholson says it even comes down to the simplest tabletop gaming experiences, such as checkers or poker. "You have a social contract when you sit down to play a game," he said. "Within this game space, we're going to do things to each other that are not acceptable in the real world. We will lie, attack, and manipulate other people. That by itself is a playful engagement. When we sit down to engage with other people over tabletop games, we're playing with each other in a very different way than we could in the real world. That's an idea that is harder to do in a digital game. Because the digital game restricts you in what you can do." With tabletop games, you are basically playing two games: the one on the table and the one around the table. With video games, you're just pressing buttons.

The best video games deliver an incredible high-fidelity experience of graphics, sound, and sensory rewards, and with the emergence of virtual reality, that will undoubtedly get better. They are unparalleled for realistic, rapid-fire action, instant gratification, and connecting disparate players over long distances. But on a social level, video games are decidedly low bandwidth compared to the experience of playing a game on a square of flat cardboard with another human being. In a digital game, even with the highest-quality webcams and microphones to capture facial expressions, we lose out on the immeasurable physical cues our body gives off . . . our posture, the sound of our breathing, the way we sip a drink, the bouncing of a leg under the table. These are the traffic lights that tell us when someone is frustrated, scared, joyous, or cocky. They are the signal flares of our most complex emotions, which shape our actions in response.

The best games engage these cues. Think about poker, and how the essence of the game comes down to suppressing and manipulating these inherent markers. Now, think about a multiplayer digital game. You are robbed of all of this, so when you shout into the microphone that another player who made a foolish move is a "stupid bitch" or some other insult, it is entirely robbed of context and consequence on your end, but not on hers. Online gaming is notorious for nasty, sexist, and downright abusive interactions between players that would get you slapped off your chair by a table mate at Snakes & Lattes. Digital players don't easily perceive one another as human beings beyond the screen. They're just avatars with guns, controlled by anonymous fingers pushing pixels.

None of this has stopped various companies, in both tabletop and digital gaming, from trying to integrate the two, with limited success. There is no shortage of unplayed board games with buttons and batteries and gizmos that sound and look interesting, but do nothing to improve gameplay. When the iPhone and then iPad came out a few years back, they were heralded by many in the tabletop industry as the future of gaming, but as Rob Daviau, a game designer (Pandemic Legacy, Funemployed) put it, iPhones on a table are a solution in search of a problem. "The reason why digital toys [and games] haven't been successful is because they suck," said Chris Byrne, author of the industry blog *Toy Time*, and one of the original creators of Pictionary. "They're just not fun. . . . You can show Wall Street that your company is on the cutting edge by merging an analog game with the iPad, but it doesn't help the actual experience." Jacobe Chrisman, who runs Wonder Forge Games, a rapidly growing Seattle company that creates board games for preschoolers based on licenses from Dr. Seuss, Disney, Curious George, and others, is highly skeptical of "tech-enabled gaming," and Wonder Forge focuses 99 percent of its efforts on analog games.

Despite this, most people in the tabletop game industry actually credit video games with the revenge of board games. In essence, all games beget gaming, and a digital gamer is more likely to become an analog gamer. "At some point it was no longer super nerdy or childish to play video games. And now everybody plays video games," said Colby Dauch, the owner of Plaid Hat Games in Ohio, which designs and sells more than $2 million in board games a year (including Dead of Winter and Summoner Wars). Plaid Hat's business has doubled or tripled each year over the past five years. "I think some of those [video] gamers think, well, maybe this love goes deeper, and they pick up a board game." Dauch also credits the normalization of geek culture as another big driver in the growth of tabletop gaming. Popular culture has fallen in love with the geek this past decade, from fantasy hits, such as the numerous Lord of the Rings movies, and explicitly nerdy shows, such as *The Big Bang Theory* and *Game of Thrones*, to *The Hunger Games* books, the scores of superhero movies, adult cartoon shows, and a newfound cool for a capella singing. Thick glasses that once would have been crushed under a jock's foot are now coveted fashion accessories, and the sexiest job title around is "computer programmer."

When Johnny O'Neal was working for Mattel around 2010, part of his job was managing its website for collectible figures (He-Man, Batman, WWE wrestlers, etc.). The average customer was a nostalgic, geeky male living in his mother's basement, the very proto-typical nerd Snakes & Lattes sought to avoid. "What's happened over the past six years is that the geeks of the world have gotten younger, and way more gender neutral," O'Neal said. "Tabletop games are a so-cial activity. It's inherently nongeeky to get together with friends and play face to face. When you start getting people in the flesh, it doesn't really work for it to be all for boys." By day, O'Neal is the director of marketing at the toy company Spin Master, in charge of toys for younger boys. By night (with his brother Chris), he runs Brotherwise Games, whose main title, Boss Monster, is a hit card game inspired by retro video games, where you play the evil boss of a dungeon trying to kill off a superhero, sort of a bizarro *Super Mario Bros.* When O'Neal first attended industry conventions such as GenCon, they were domi-nated by men. Today, he estimates, half the attendees are female.

That move from the dark corners of nerdism into a friendlier geekdom has pushed the tabletop game industry into the mainstream. As I spoke with O'Neal by phone as he was shopping in a Target store near his home in Los Angeles, he remarked on the slow embrace of the potential for board games by the larger toy market. "Look, here I am staring at fourteen games on the shelf here in Target that truly would have only been found in hobby stores up until a year ago," he said. These included Settlers of Catan, Ticket to Ride, Munchkin, Pandemic, King of Tokyo, Small World, Alchemist, The Resistance, and even a new edition of Dungeons and Dragons. All in Target, America's store, sharing space with Scrabble and Trivial Pursuit, America's games.

*I*t is one thing to make board games more appealing to the mainstream. It is another thing to get people to play those games and enjoy them. At Snakes & Lattes, this is where game gurus come in. A game guru is the equivalent of a sommelier at Snakes & Lattes, a knowledgeable, customer-driven guide to the three thousand games the café has in stock, whose job is to ascertain a table's taste and provide just the right game for the occasion. The café's head guru is Steve Tassie, a middle-aged, ponytailed gentleman with a narrator's booming voice, who wears a different Hawaiian shirt each day. A stand-up comedian, teacher, board game designer, and veteran of game and hobby stores, Tassie was offered a job by Castanie within six months of Snakes & Lattes' opening, because he was its most loyal and knowledgeable customer. Since the café's customers are split among total gaming neophytes, moderate gamers, and a small but vocal percentage of hard-core gamers, a guru's key skill is reading a table's social cues. Because most customers don't know what they want, and usually ask for "something fun," it requires a quick mind, encyclopedic knowledge of games, and the ability to sell something in a twenty-second pitch. It is the very same, deeply human ability required to win at gaming, which is why gamers make such good gurus.

"We need to have the ability to be passionate about games without being preachy," said Tassie, whose shirt was patterned with flaming Polynesian masks, the first time I met him. "It is the opposite of the board game store. It's not our job to put down Monopoly. We are ambassadors to the hobby. We are endeavoring to give people fun, but also change their concept of what fun can be. A lot of games people liked as kids have probably been visited by the Suck Fairy." Tassie had actually heard people request the classic The Game of Life, play it for ten minutes, and ask for something else, because, as he puts it, "'Life' sucks." Instead, a guru's job is to take a request for a tired game and turn it into an opportunity to introduce a new game. Ask for Monopoly and Tassie will suggest For Sale or I'm the Boss, both financial games with faster, more conclusive gameplay than Monopoly.

I tested out Tassie's skills one day with my friend Wendy.

"What types of games do you like to play?" Tassie asked us.

I told him trivia games were my strength, and that we wanted to try a few games but only had an hour to play. "Whatever I won't suck at," Wendy said with a smile.

He brought over a small box and launched into an explanation that was so clear and confident, he could have been selling a car. "This is Timeline," he said, "a chronological trivia card game. Each card represents things created: movies, albums, books, songs. The date of their creation is on the back. Your first card is the center of the timeline, and your goal is to get rid of your hand by correctly placing cards in the right place on the timeline, before revealing the date. If you place one wrong, you have to take another card."

Our first timeline began with James Brown's "Get on Up/Sex Machine" (1970). We quickly threw down cards onto the timeline: *Rebel Without a Cause, Sunday Bloody Sunday, Texas Chainsaw Massacre.* Wendy won the round by placing *2001: A Space Odyssey* between *The Bridge on the River Kwai* and *Taxi Driver.*

"Wendy, showing up!" she yelled triumphantly. Two more rounds of Timeline and we asked Tassie for something different.

"Now we're getting into the world of abstract strategy," he said, opening up Quarto. "This is tic-tac-toe's sexy European cousin. You need to connect four pieces in a line, but your opponent chooses which piece you have to play." Quarto looked deceptively simple—a classic game that is played on a grid with differently toned and shaped wood pieces—but within two rounds, the play became intense.

"Looks like this round's mine," I smugly told Wendy, as I prepared to place my winning piece on the board.

"Really?" she asked even more smugly, pointing out a perfect diagonal line of rounded pieces she had just laid down. "So, that doesn't count?"

Damn it!

Finally, Tassie brought us Jaipur, a card-trading game where your goal is to get rich in an Indian marketplace, which involved a lot of buying, selling, and hoarding camels. It was a bit complicated to figure out, but within a few minutes, the camel jokes were flying and our fortunes at the bazaar piled up.

Tassie manages a staff of half a dozen gurus, visits gaming conventions to source new games, and creates regular videos, blogs, reviews, and podcasts for Snakes & Lattes' website. But his most important role is the curator of the café's games library, its core asset, which is constantly being edited. One day, I joined Tassie (in a shirt printed with vintage Cadillacs) and Todd Campbell, another guru, for a test play of new games. The upside of the boom in the board game market is that there is a constant influx of new titles for Snakes & Lattes to feature in the café and sell at its retail store and online. But with thousands of new games published a year, there is also a huge variance in quality, and everything that gets featured at the café must be play tested by Tassie and the gurus. Tassie took several boxes to a table and set them down.

The first one, Say Cheese!—a selfie-themed matching card game from Taiwan—was quickly discarded for its strange art, including what Tassie deemed a "weird insistence on bunny rabbits" and borderline racist depictions of minorities. Next up was Red 7, a simple-looking card game of numbers and colors that turned out to be more complicated than it should have. Before they had even finished unpacking What the Food?!—a food-fight card game—Tassie and Campbell were putting it back in the box without even checking the rules. It had way too many small, easily lost pieces for a game with a fun, easy-sounding premise. "I'm a curator," Tassie said. "I'm there to make sure stuff on the wall is there for a reason: good games that should be getting played." The last game was a small one called Samurai Spirit, where seven different samurai warriors with magic powers defend a village against ghosts. It is what's known as a cooperative game, where players work together, rather than against one another, to achieve a common goal.

Good games are difficult to create, and Tassie primarily credits the renewed interest in board games to a shift in game design. Previously, games were split into two camps: "Ameritrash versus Eurogames," Tassie said with a smirk. Historically American games tended to be simpler in their play, and yet they would go on for far too long, with stretches of inaction for most players and no definite conclusion. When was the last time you actually finished a full game of Monopoly? Exactly. European games were heavier on the strategy, but they were so rule intensive you could hardly play without contacting your lawyer. Also they were dry, both in content and design, especially next to the great, poppy graphics and characters of American games.

"Over the past five to ten years, there emerged a real hybridization between the continents," Tassie said. The result is a new breed of designer games that look American in their production values, but play with a European sensibility. These games are strategic, often cooperative, and inventive. They take less time to get started and have definitive conclusions, with a victor emerging in an hour or less. Many in the industry say that we are living in a golden age of game design, and most credit the emergence of this to the success of Settlers of Catan, a game invented by a German dental technician named Klaus Teuber.

Originally released in Germany in 1995, Settlers of Catan has gone on to sell more than 23 million sets, including various editions in over thirty-five languages. The game tasks players with settling the island of Catan, the topography of which changes with each game. Players gather resources and sell or trade them to finance settlements and roads, until someone achieves victory. It stands out because the board is always different, the rules are clear, and the game is played in under an hour. Yet the capacity for deploying strategy, by building alliances or using emotions to foster deception, is limitless.

Settlers of Catan was the first big European crossover tabletop hit in America. It initially gained a foothold in colleges and the technology industry (Facebook CEO Mark Zuckerberg is reportedly a huge fan), but as its cult following turned mainstream, it woke the nation up to this new concept of play. "The rise of Eurogames parallels the rise of digital games, but Eurogames are insistently analog," Adrienne Raphel wrote in the *New Yorker*, in a profile of Teuber. "They can be as complex as video games, but, because there's no fixed narrative, groups of people play together over and over." *Wired* magazine and the *Wall Street Journal* hailed it as the successor to Risk and Monopoly.

Klaus Teuber's son Guido, who helped his father create Settlers of Catan and runs the US division of the company from Oakland,

California, says the game's success initially surprised his family. "On the face of it, it was boring," he said, "but there was something beyond that. When you explain the rules, people yawn, but once you do your first play, there's a spark. We didn't realize it had the potential to go beyond the core gamers. Even today, when I explain the game, people yawn. But once they start, it triggers something." This certainly mirrored my first Settlers of Catan experience, after I bought a copy at Snakes & Lattes to play with my wife and our two friends. It seemed as if we were reading rules and staring at cardboard pieces for half an hour, until our friend Vanessa, who had played it before, urged us to start. Once we did, a fierce colonialist landgrab ensued, with cutthroat construction, merciless trading, and deception galore. "Wow," my wife said, when she suddenly claimed victory, "that was awesome."

There are digital versions of Settlers of Catan, but they have not proved popular in nearly the same way as the original, cardboard game. A lot of that comes down to the fundamental role of tabletop games as an excuse for human interactions. "When people want to have the analog experience, they *really* want the analog experience," Guido Teuber told me. "Yes, there's skill, there's luck, but also the ability to communicate. . . . If you use emotional expressions, you can be successful at the game." This cuts across lines of gender, class, and culture. Although Catan is played around the world, in extroverted cultures (Brazil, Italy, Israel) and introverted ones (Germany, Japan, England), Teuber says the game creates a level playing field, which gives permission for open people to be more guarded, and soft-spoken players to get aggressive. To win at Catan requires strategy, but also a high degree of emotional intelligence and instinct. That is the key to its success in an entertainment market still dominated by digital games. "Settlers of Catan is about negotiation and organic, woolen concepts; bluffing, lying, deception, etc.," said Paul Dean, a former professional video game critic who cofounded the popular tabletop gaming blog *Shut Up and Sit Down*. "A computer can't do that well. A computer can do chess well."

Arguably, Catan didn't even start the tabletop comeback. Ticket to Ride, a railway-building strategy game produced by Days of

Wonder, a company founded by two successful Silicon Valley entre-preneurs, preceded Catan's American success by several years. Today, there are lots of new titles in this category, from the cooperative medical crisis game Pandemic to games that are based on movies and TV shows, such as *The Walking Dead* and *X-Files*. Not all these new games require heavy bouts of strategy. They also include fun party games; for example, Kwizniac and Hey, That's My Fish!

Perhaps the most significant factor behind the tabletop game boom is the way that digital tools opened up tabletop game design from a closed industry to one brimming with new entrepreneurs. First came the online community Board Game Geek, created in 2000, which provided a forum for tabletop lovers to discover new games, share ideas and advice on design and commercialization, and orga-nize in-person gatherings. YouTube, Facebook, and Twitter expanded on this with fan pages, news about games, and game review videos. The best-known voice in the industry belongs to Wil Wheaton, the actor who played Wesley Crusher on *Star Trek: The Next Generation*, and whose hilarious YouTube board game review show, *Tabletop*, can clock over a million views per episode. The show is so popular that game industry insiders refer to the sharp spike in sales a game receives after it gets featured on the show as the "*Tabletop* effect."

Other digital tools include low-cost, open-source game design software and templates, 3-D printers (it's a lot easier to print a small dragon than carve one by hand), and print-on-demand services, such as DriveThruCards, which lets designers upload a card-based game, printing off each order as it comes in. By far the most disruptive and powerful technological tool behind the revenge of tabletop games has been Kickstarter. Since the crowdfunding service began in 2009, it has quickly become the de facto launchpad for tens of thousands of board and card games, large and small. At any given time there are roughly two hundred new tabletop game projects raising money on Kickstarter, and roughly half reach their fund-raising goal. Tabletop games are one of the most popular projects on Kickstarter, in terms of both dollars raised and the success of fund-raising campaigns. Kick-starter does not regularly break down its statistics for the games cate-gory (which includes both video and tabletop games), but in 2013 the company told the *New York Times* that tabletop game projects raised $52.1 million that year, compared to $45.3 million for video games.

Kickstarter has done more to fuel the creation of games than any-one since Milton Bradley. Almost every single designer I spoke with for this book had launched games on Kickstarter. There are certainly runaway successes, such as the silly card game Exploding Kittens, which raised over $8 million in a matter of days, but most projects raise a few thousand dollars to pay for a game's production. Some games start out small on Kickstarter, and eventually grow huge. One of the first to do this was Cards Against Humanity.

Cards Against Humanity is a game that asks players to complete a sentence, such as "This season, Tim Allen must overcome his fear of _____ to save Christmas." with choices that include "Anal fis-sures like you wouldn't believe," "Ribs so good they transcend race and class," "The shambling corpse of Larry King," and "The Jews." The goal is to complete the sentence with the most offensive card, the-oretically gaining points. But no one really keeps score when playing Cards Against Humanity, especially because the players tend to be drunk. There are no winners to the game—just the gradual erosion of polite society, wearing down, as one of the game's answers puts it, like "the tiny calloused hands of the Chinese children that made this card."

Cards Against Humanity is the cocreation of eight childhood friends from the Chicago suburbs, who invented it at a New Year's party while home from college and made it available for free online. Over the next two years it gained a cult following, and its creators turned to Kickstarter in 2011 to see whether they could get it printed. "Kickstarter gave us an incredible opportunity," said Max Temkin, one of the game's creators, who manages the business part-time. "We had an idea we wanted to make and weren't sure other people thought it'd be funny. We also needed $4,000 to do the first game, and print it at a print shop near me here in Chicago. We didn't have $4,000. It was our only option." In the end, they raised more than $15,000.

In print, Cards Against Humanity subsequently took off be-yond all expectations. The company does not disclose sales figures, but some estimates have put the number of sets sold (with an average price of $25) at more than a million. It is consistently the top-selling game on Amazon, and makes up 60 percent of games sold at Snakes & Lattes, its Canadian distributor. "Prior to Cards Against Humanity, we were a tiny little operation," said Aaron Zack, who helps run Snakes & Lattes' online retail and distribution business. "Now we have our own postal tier, thanks to them."

Temkin and his cohorts aren't above using their success to make a point about the excessive culture of consumerism that fuels it. For Black Friday 2013, they sold the game for $5 more than usual, advertising this price gouging explicitly. The overpriced sets sold more than the previous year's. The next Black Friday, they removed the game from the online store, and promised to sell their customers a box of "bullshit" for $6 instead. Thirty thousand eager fans scooped up this deal in under thirty minutes. Weeks later, each received a box of actual dried bull's shit in the mail, with a note stating that the proceeds were donated to a foundation supporting financial transparency in American politics.

Cards Against Humanity's success has certainly made it a target for critics within the industry, who object not only to its offensive content, but to the way it gets lumped in with other, more sophisticated games. Gamers feel it debases the entire gaming experience. Instead of being an entry point for a hobby people can grow to love, it is a crude, one-off gimmick, no more likely to get a neophyte into tabletop gaming than a porn film will get its viewer into cinema. Others say the gameplay itself isn't nearly as clever or funny as its reputation would have you believe.

These criticisms are fair, but they miss a key point about Cards Against Humanity: it was never created to be part of the great movement of designer tabletop games. It was the ultimate game as social lubricant, so stupidly simple, ridiculous, and juvenile that any group with a strong enough constitution can pick it up and start laughing in seconds. Although it sits on the entirely opposite end of the spectrum from Settlers of Catan, it equally distills the appeal of the analog gaming experience down to its essence: human contact.

"We're a very lonely generation," Temkin, who is twenty-nine, told me. He and his friends who created the game (all card-carrying nerds) grew up on the Internet. In high school, they hung out in chat rooms and played *Starcraft* online. By the time they were in college, they hungered for real connections. "There is a tension between the ubiquity of tools of connection, which share carefully curated images of our lives, and the reality of our own lives not meeting up to that on a real basis. You have a million friends on Facebook, but none to hang out with in real life. How perfect does Instagram look compared to the sad reality of your life? Once you start feeling that loneliness, you'll turn more to social media and risk-free forms of connection, looking for a quick fix, going back, clicking and refreshing that feed again and again. Everyone has that experience of getting into that cycle of newness and trying to find a connection."

Temkin called digital social media the anesthetic of loneliness, paraphrasing David Foster Wallace. "That's why it feels so good to sit down at a table with friends in real life and play this game," Temkin said. "You do it and the connection is immediate, you know what the rules are, and it's an easy social relationship to navigate in our context. Cards Against Humanity was made by us as friends to socialize and laugh together. Cards Against Humanity answered that need at the right time. It was a great excuse to have a gameful, safe way to push boundaries of taste. My hope is that it led people to real moments and connections together."

Many of the same critics of Cards Against Humanity also believe that Kickstarter is bad for the tabletop game industry. They see easy crowdfunding money flooding the market with poorly designed games, projects that never get made, or games delivered long past their deadline. These not only turn people off tabletop games, the critics say; they saturate a relatively small market with more product than its consumers can healthily support. Perhaps there is some merit to that, but the alternative—creating a game from spec, spending your own money on a prototype, having it tested at various game and hobby shops, and then trying to sell it to a traditional publisher at a convention—is not a process game designers are eager to return to.

*O*ne *night, I walked over* to Snakes & Lattes to meet Alejandro Vernaza, who was in the final week of a Kickstarter campaign for the game Deal: American Dream. Vernaza is originally from Bogotá, Colombia, but now works as a teacher in Toronto. The game is a coventure between himself and a team in France, including Tristan Frobert, with whom Vernaza shared a cramped train cabin on a six-day voyage across Russia in 2010. During the trip, they quickly grew bored and began making games out of paper, but shortly after the trip, Frobert wrote to Vernaza and told him about an idea he had for a Risk-style game based on drug cartels. Over the next few years, they worked to develop the game, and took early prototypes to board game industry trade shows, including Germany's massive Essen Spiel, where they learned about Kickstarter. When they launched their Kickstarter campaign in late May 2015, the goal was to raise €29,000 to fund Deal: American Dream. I met Vernaza with just six days left, and only €20,000 raised.

"No one tells you how much Kickstarter is a ride," he said, as he unrolled a prototype board for the game and set up the cards. "It's really a ride."

Deal: American Dream sets competing criminal networks against each other in the drug-producing and -consuming markets of the Americas. You pick a gang (Chicago mafia, Mexican cartels, Vancouver yakuza, etc.), raise funds, buy soldiers, take over territory, then move and sell dope to earn respect points. The first player with ten respect points wins the game, but to get there you have to cope with competitors trying to fleece you, invade your turf, and kill you with all sorts of weapons. As we kicked off play, with Notorious B.I.G. appropriately playing on the café's stereo, I consolidated territory along the US East Coast. Vernaza took the Southwest, and I fended off his attacks in the Midwest and Texas until I controlled Miami, and finally his home nation, Colombia, where I shipped enough dope back to America to win. Deal: American Dream was a fun, fast-paced, original game. I pledged $10 to the campaign, and followed Kickstarter over the next week. With five days to go, it looked increasingly unlikely that Vernaza and Frobert would reach their goal, but a sudden surge of backers in the last three days pushed them past their goal. The game should now be commercially available.

Deal: American Dream had filmed its Kickstarter video at Snakes & Lattes, and I first encountered Vernaza there just before it launched, during one of the monthly game designer nights the café holds in its back room.

If Kickstarter and Board Game Geek are the digital communities driving the revenge of board games, then these evenings are their analog equivalent. Each month, twenty to thirty game designers, ranging from well-known professionals to first-time amateurs, invite their peers to play and give feedback on prototype games. The crowd is less diverse than Snakes & Lattes normally. It is more male and geeky (Star Wars T-shirts abound), but the breadth and scope of the games on display are truly representative of the creativity this new age of game design has unleashed. There were simple card games about bike couriers and home contractors, tricky-looking number games, funny trivia games, games that were written on cut-up scraps of paper, hand-carved wooden boards with intricate player pieces, and slick-looking games that looked store ready. I saw a game based on the tale of the Pied Piper and another on postal carrier pigeons crossing the English Channel in World War II, and many, many fantasy games.

Tassie was there in a shirt printed with tropical cocktails to provide aspiring and seasoned designers with advice and feedback, but also to test out a redesign of his B-movie card game Grave Robbers from Outer Space, which first came out in 2001, but was no longer in print. He was considering going to Kickstarter with an updated version. "I think there's a market," Tassie said, laying out different monster, prop, and character cards (including "Guy Who Will Die First" and "Girl Who Shows Her Breasts"), which players formed to make a movie. "It's my first Kickstarter campaign, and it's pretty fucking scary. I already know how much fun the original game is to play, but what I'm looking to do is show it to people who have never played it before, to ensure that the fun is still there." Kickstarter and Board Game Geek were great tools for fund-raising and publicity, Tassie said, but the only way you can truly get a sense of a game's actual chance of success is to bring it to a place like this and put it in the hands of people who have never played it before.

One of the designers present that night was Jenny Mitchell, a quiet artist, musician, and schoolbus driver from outside Toronto, who was testing out her first game. Mitchell was into crafts and repurposing old finds. She and her friends regularly got together throughout the winter to play board games they salvaged from secondhand shops. "My game is called Hoarders," Mitchell said, "and I am a hoarder." Mitchell's father had owned an antique store and regularly encountered hoarders. "One woman died standing up," she told me. The game was kind of her way of coping with a slightly perverse habit she'd picked up, without the risk of perishing upright amid stacks of yellowing newspapers.

Hoarders was crude but effective: a mix of glued cardboard, handwritten rules, dollar store money, and tiny cards Mitchell had painstakingly illustrated. To play, you scoured the dump for useful scraps (wood, metal, wires, tools), then combined those to make saleable items, such as a radio, a table, or a furniture set, which you sold back to the dump or other players. The rounds were counted in weeks, and every week brought garbage day, as well as flash floods, health inspections, birthdays, garage sales, and tornados, which could either hurt or help your hoarding. The player with the most money after twelve weeks won.

I played with a couple named Lex and Terry (who nicknamed me "Paparazzo") and François, an older designer with a wizard's beard, who asked Mitchell whether there was any particular feedback she was looking for. "Probably around the rules," she responded, "but I hope the natural order will come out." We set up the board and got going under Mitchell's guidance. At first everyone acquired what they could, until the selling began with a few small items. During week five, after a tornado shuffled up our cards, I saw a stray metal piece and swiped it from Terry, who shouted, "Paparazzo!!!" in mock rage. As the game went on, the tricky areas of play came out, and suggestions began flowing to Mitchell. Things were slow between turns, so François suggested allowing players to trade at all times, not just on their own turn. "Maybe there's a deck of cards with cats on it, and the winner is showered in cats," Terry said, shortly before finishing second behind François.

Toward the end of the night I stood over a table in the corner, where five men were playing a kids' game called River Runner. Players had to match cards by memory so as to cross a raging river. It looked so simple and perfect, I asked the designer, Joshua Cappel, whether I could buy the prototype right there. These guys were members of the Game Artisans of Canada, a community built around mentoring and helping fellow designers get their games to market. It was partly led by Sen-Foong Lim, an occupational therapist and professor of developmental psychology, who had designed games as diverse as the hilarious infomercial party pitch But Wait, There's More! and an adaptation of the TV show *Orphan Black*.

Lim saw game designing as a "jobby," a mixture of a *job* and a *hobby* that made just enough money to justify the time it required away from his family. Even though there were blockbuster games out there, the vast majority of tabletop games were made for the love of gaming. "My goal is simple," Lim said. "The more games we put out there, the more the culture of gaming grows." Lim acknowledged the power of digital tools to bring that analog culture to life. He had raised money on Kickstarter, was a regular on Board Game Geek, and even hosted his own podcast. But he also firmly believed that the revenge of board games was due to the real, physical communities made possible by such places as Snakes & Lattes.

"Snakes is proof," he said, looking around the café, which was completely full on a Monday night at 11:00 p.m. with people of all ages, sexes, backgrounds, and interests playing all sorts of games. "Proof that this hobby is further reaching than a bunch of guys living in their mom's basement, playing D&D. This is a real place where real people play *real games*," said Lim, as he set up yet another game and his friends all gathered round to play.

PART II

THE REVENGE OF ANALOG IDEAS

5

The Revenge of Print

Shortly after my last book, The Tastemakers, *was published, a woman* at a prominent New York marketing agency contacted me. She had read the book, and wanted to discuss an online show about food trends they were developing for a client. We met for coffee in New York. Our conversation quickly turned to my current work, and I told her about *The Revenge of Analog*. She thought it was an interesting idea, but had a question.

"Why a book?" she asked.

What did she mean?

"Why would you write another book?"

She then proceeded to spell out all the arguments against publishing a book as a means of expression. Books took tremendous amounts of research and writing, publishers paid relatively meager amounts for that work, and very few books sold well enough to make their authors any significant money. Besides, no one read print anyway.

What was the alternative? I asked.

"Branded content," she replied, as though it was obvious. The world was driven by brands, she said, and artists and other creatives were producing great works on behalf of brands. She could easily see *The Revenge of Analog* as one of these projects. Perhaps Sony Music would pay for a web video series about the vinyl record industry, or Canon would sponsor a blog about analog "makers." People didn't want more books, she told me. "They want digestible bits of branded content," and those like me who create art or ideas can either make money producing that content or grow poor while we desperately cling to our dying totems of culture.

I left the meeting profoundly shaken. Who the hell did she think she was, this former model with a nose like a ski jump, to question my very professional existence five minutes after we met? As days went by, her question ate at me. Was I so upset because she was right? Why the hell *was* I writing another book?

Since I began writing professionally in 2002, the specter of print's death at the hands of digital has been so consistently present that I came to accept that truth as the natural order. Working in printed media (books, magazines, and newspapers) feels similar to life in a rustbelt city, where you take comfort in the past's fading glory as the world contracts around you. Each year that I have made my living as a writer, more publications I have written for have folded, more magazines and newspapers have seen their pages cut, more editors have been laid off, and less money has been made available for the work I was producing. Print publications seemed to be moving in one direction . . . down . . . pulled there by the inexorable law of digital gravity.

Print publications cost more to produce and distribute than digital ones. To deliver a magazine, newspaper, or book to a reader requires trees, paper mills, giant printing presses, fleets of trucks, warehouses, postal carriers, and retail stores. Print publications cost money to buy, and they take up a lot of space. By contrast, a digital publication requires no physical resources, no humans to deliver, and takes up no space. A digital publication creates no waste, and it is either free or absurdly cheap. You can produce and distribute one copy with the same ease and cost as one million.

Knowing all that, why would anyone choose print?

And yet, print has not only endured, but in certain areas it is growing, spawning new publications, or even new analog versions of publications that began online. And while few businesses are making billions in periodicals right now, it has become increasingly clear that digital publishing's triumphant narrative about low cost and instant delivery as the key to success isn't the whole story. Analog solves all kinds of problems that are albatrosses to digital publishers—engagement, stickiness, discovery, etc.—and, if the order of inventions were somehow reversed, print could easily be presented as the truly disruptive technology.

A *few months after that awkward* coffee, I found myself in the packed basement of a bar called the Book Club, in London's trendy Shoreditch neighborhood. The city, and specifically this part of its East End, was the center of a vibrant new print publishing scene, focused largely on magazines. A sold-out crowd of nearly a hundred people in their twenties and early thirties gathered under a ceiling covered in thousands of spent lightbulbs for an event called Stack Live. Steven Watson, creator of the independent magazine subscription service Stack, hosted a monthly Q&A with the creator of whatever magazine Stack distributed that month. Tonight, Watson was interviewing Rosa Park, the extremely verbose American cocreator of the biannual *Cereal*, which was a fast-rising star of the country's independent magazine publishing scene.

Park was born in Seoul, grew up in Vancouver, and worked in fashion marketing in New York before she moved to Bath, England, for a master's degree. There she met Rich Stapleton, a British designer, and in 2012 the couple launched *Cereal*, a design-focused travel magazine with a sparse, Scandinavian aesthetic. Their first print run was for 1,500 copies. In contrast, their 2015 fall issue printed 35,000 copies, which quickly sold out, as all *Cereal* issues now do (secondhand copies sell online for double their cover price or more). Last year, *Cereal* launched editions in Korean, Chinese, and Japanese; increased its distribution in the United States; created a separate literary supplement; and launched collaborations that have included furniture, paintings, and a line of ceramics. Presently, Park was setting up a vinyl record press, as well as a line of flower arrangements. Park and Stapleton did all of this with just two full-time employees.

"I don't understand," Watson said, with a laugh. "You must not have kids."

He asked Park about *Cereal*'s newly launched city guides, and why she and Stapleton decided to release them in print rather than digital, given the financial and logistical hurdles of working in ink and paper. My ears perked up. "I was actually adamant that it would be online originally," Park said. "But the request for print was simply overpowering. We printed our London guide as an experiment and it sold out in two weeks. Now, we've shifted the model to producing a paid printed version, and a smaller, condensed online one."

I raised my hand and asked Park what advantage she felt a print magazine had in a market where it competed with digital titles. "People will get to know a print product much faster than they will online," she said. "Because our product is in your face when you walk into a store, a hotel, or when someone is reading it next to you on the train. Once you get your product out there, someone will get it noticed." As Park spoke about her vision for *Cereal*, the magazine's finances, and why they never gave away a single copy for free, I looked around and saw many in the audience with their heads bowed, carefully scribbling her comments in notebooks. These were not only *Cereal*'s fans, but the people driving the revenge of print. They were all young, digitally savvy, and yet here they were, paying for the privilege of Park's take on *Cereal*'s success in paper, which they hoped to replicate with their own magazines.

"The headline 'Digital Will Kill Print' is a very simple, convincing narrative that's not true," said Jeremy Leslie, the morning after the Stack Live event, as we chatted around the corner over tea. A former designer for newspapers and magazines, Leslie has written two books on magazine design, and runs the popular blog *Mag Culture*, which tracks the industry worldwide. "When someone says, 'Tell me why it isn't the end of print,' I tell them that despite all the problems, any publisher of any scale still makes the bulk of their money from print, and there are more titles than ever before."

The magazine industry has always been defined by churn, Leslie said. For each publication that closes down and triggers further obituaries for print, a great many more new magazines are born. According to statistics compiled by Samir Husni, an American academic who goes by the moniker "Mr. Magazine," on average twenty new regular-circulation magazines are launched each month in the United States alone. These include trendy magazines, such as *Cereal*, *Little*

Brother (a micro-magazine of essays), *Kinfolk* (home for hipsters), *Lucky Peach* (global food culture), *Drift* (coffee and travel), and *California Sunday* (the *New Yorker* of the West Coast), as well as titles ranging from *Gay Wedding Magazine* to *Girls Guns and Rods*, *3D Make and Print*, and *Guinea Pig* (about furry animals, not human experiments).

What these titles demonstrate is a shift toward a new model of producing and selling print publications, specifically designed to work in a postdigital economy. The new crop of magazines have largely started small, growing their audiences organically, with initial print runs of just a few hundred or thousand copies. That's the opposite of a large launch from a legacy publisher, which might hit stores with hundreds of thousands of copies of its first issue. Yet many of the independent magazines that have launched over the past decade are now approaching circulation levels similar to magazines produced by multinational publishing companies. "What we're going to start seeing is big publishing companies like Condé Nast and Hearst emulating the model of the independent publishers," Leslie said. That meant smaller print runs of a higher-quality product, targeting a smaller, more valuable audience of readers.

Leslie credited a substantial part of this new global class of print magazines to digital technology, especially the proliferation of desktop publishing software, which allowed any size publisher to produce a great-looking magazine. Steven Watson agreed that this renaissance in print had its roots in digital. "There has been this boom in magazines," Watson told me when we spoke at the Book Club shortly before the Stack Live event with *Cereal*. "Not more money for magazines or more magazines being sold, but more magazines starting, thanks to technology." These included design software, crowdfunding campaigns, and digital printing, but Watson also felt a lot of the credit went to blogging. "People blogged, and they grew up with blogs and writing as a form of self-expression," he said. "Now, those same people want to be in print, so their ideas are legitimate." The permanence of paper conferred another level of credibility that just wasn't achievable online.

As Watson saw these new magazines grow, he noticed a fundamental gap in the market. There were lots of new magazines, and lots of fans for these magazines, but the system of distribution was geared toward big publishers. The magazines published by these companies (*Time*, *Playboy*, *Golf Digest*, *Forbes*, etc.) were printed in huge quantities, sold at a loss, and entirely subsidized by advertising. "The traditional magazine market is based on a dreadfully wasteful model, where you print twice what you need, pump them into spaces, and pulp the half you don't sell," Jeremy Leslie said. The newer niche-focused magazines that Watson saw emerging, such as the innovative German publication *MC1R*, which is a lifestyle magazine for redheads, could never succeed that way. It was simply too costly. This is where Stack comes in. Each month, Watson picks a different independent magazine and sends the latest copy of it to Stack's subscribers. September could be *Address* (literary fashion); October, *Elephant* (arts and culture); November, *Intern* (for the unpaid laborer); and December, *Wrap* (graphic design, with wrapping paper inserts). Stack subscriptions cost $190 a year.

Watson began Stack as a part-time project back in 2008, when he was still working as an editor, and by 2014 he had acquired more than four thousand subscribers. He hopes Stack will be up to ten thousand subscribers by 2018. The business (which is slightly profitable) makes money by buying these magazines at wholesale prices. Magazines love Stack because it guarantees sales and brings in new readers, and readers love Stack because it introduces them to the best new magazines, at a discount from the cover price. Watson is eagerly expanding Stack into North America, and launching new distribution services that include the ability for magazines to sell readers single issues and annual subscriptions through Stack's website.

"Steve Watson is an absolute legend," said Rob Orchard, the cofounder and editor of *Delayed Gratification*, a magazine that was Stack's pick for January 2014. *Delayed Gratification* tackles the news with a retrospective, analytical take they call slow journalism, sort of an antithesis to the insta-punditry of digital publishers. "We launched around the same time, and we've always supported one another. Stack has been absolutely fantastic in generating interest in magazines. It is a very practical use, because you can have a small print run and he'll give you a guaranteed audience, and paid at that. He's a very smart character. He's looking at all problems facing small magazines."

One of the unifying things I found while speaking with people in the United Kingdom's independent magazine industry is their sheer bullishness on print and its advantages over digital publishing. This was not a romantic notion but a firm economic argument. "The key one is that people will pay for print," Orchard said. Print, he continued, is a tried-and-true business model. *Delayed Gratification* takes no advertising revenue, sells roughly five thousand copies of each issue, publishes four times a year, and grosses more than £200,000 in annual revenue. He does this by selling the magazine for more than it costs to produce. That may not seem like much money, but it is more economically viable over the long term than a publication that is shedding millions while figuring out its business model. "I still haven't seen a truly successful digital publication," Orchard said. "I've seen lots of people dive into digital, thinking it's the answer, and then not know how to make money."

———————

While digital has obvious advantages in its distribution, the profit model for digital publication remains largely uncertain. For all the bravado about the death of print, most digital publications still spend more than they make. That cuts against the core assumption of the digital publishing industry, which was summed up for me back in October 2008 by the comment someone made to me at a party thrown in New York by Gawker, the media gossip website-turned-digital publishing empire, where this person worked as a blogger. Lehman Brothers had collapsed weeks before, the stock market was in free fall, and the recession was bearing down on the media industry. Print publications were closing and shedding jobs. "This is fantastic," she said of the doom and gloom she was chronicling. "All the advertising is going to flee Condé Nast and finally come to us!" She was half right.

The recession accelerated the shift of advertising money from print publications to online sources, but unfortunately, it did not migrate dollar for dollar to digital publications. Instead, the money spread out all over the web: Craigslist and eBay got the classifieds; Google Ad words and Yelp got the local ads; and other brands and advertisers paid for shares of page views in certain categories that are automatically filled by software, wherever your browser lands. The same pool of advertiser money went to more places, driving down the cost of ads and making both digital and print publishers poorer in the process.

One day in London, I spoke with Jane Wolfson and Steve Hare, who work for the media-planning agency Initiative, which coordinates advertising purchases for such brands as Coca-Cola and Amazon. Hare said that the trend away from print had decreased its share of media spends from 30 percent in 2005 to just 7 percent in 2015, with that lost share going to digital (TV, billboards, and other media have remained steady). "But I think that's a symptom of headlines," Wolfson said, noting that clients were reluctant to invest in something they were repeatedly told was dying, even though print's strengths to advertisers had not actually diminished. "Print has always performed well on pure ROI [return on investment]," she said.

Print ads have a higher engagement than digital ads. They are seen for longer periods of time, and both the brand and publication have far greater control about where and how an advertisement is presented, compared with algorithmically placed ads in digital publications (which might display an ad for a fake Kate Middleton sex video on the website of the *New York Times*, to cite a recent example that graced my computer screen). While digital advertisements are often seen as intrusive nuisances and obstacles, begging to be clicked away or prevented altogether with ad-blocker software, paper ads fit into the editorial landscape. In the case of a publication such as *Vogue*, they are as much a reason to buy the magazine as its editorial content.

"I would say a print reader is worth more than a digital reader," Wolfson said. "They have a higher affinity and loyalty to that product [the magazine or newspaper], that transfers to brand advertising." According to a recent survey by Magnetic, the marketing agency of the British magazine publishing industry, 90 percent of magazine readers look at advertisements (far higher than any other media), and 70 percent of participants bought something or visited a business after seeing an ad in a magazine. Online that number is significantly lower. "It is very difficult to build a brand online," Hare said, "because you click in, click out, and it's very difficult to build a relationship with someone in that."

All of this translates to the revenues publishers are able to gather from advertisers. "Digital advertising is still twenty cents on the dollar, where a magazine ad is one dollar," Nicole Vogel told me, when we met in 2014 at her office in Houston, Texas. "Digital ad revenue is just the gravy." Vogel is the president and founder of the city magazine *Houstonia*, which is part of the Saga City Media company she began with her brother a decade ago. Saga City now owns more than fifty paid monthly city magazines across the United States, including *Portland Monthly* and *Seattle Met*, and all these magazines have growing print subscriptions, which remain the core of Saga City's business. "Print is great for delivering a marketing experience," Vogel said, "because it's uninterrupted and you decide. And because you decide, you'll give it more time than if it has chosen you, like a pop-up ad online driven by an algorithm. When that pops up, you're immediately looking for the X to close it, but when a print ad comes, you aren't ripping it off the page because we forced you to see it."

Strip away the venture-capital backing most new digital publishers exist on, and you have companies paying money to produce content for more than they are able to generate in ad revenues. That isn't a revolutionary change in publishing or a fix to a broken industry, it's just bad business. I have been approached countless times by editors from blogs, apps, websites, and other "new forms of digital storytelling" with requests to write for them. When I ask what they can pay, the answer is usually some variation of "We can't afford to pay writers at this time, but you will get great exposure." I always tell them that I have all the exposure I need right now. What I'd really like is money. "Digital pennies versus print pounds," a writer in London said, characterizing this race to the bottom.

One of the big differences between print and digital publishing is the ability to charge for the product at hand. Most print publications cost a few dollars to purchase, while digital ones are free. This might seem great in terms of capturing as many readers as possible for advertisers, but what it actually does is drive down the value of those readers. Most new independent print magazines opt to sell their issues rather than give them away, because the free-content model is deeply troubled, even digitally. "Online media is now drowning in *free*," media journalist Michael Wolff wrote in a *New York Times* op-ed in 2015. "Google and Facebook, the universal aggregators, control the traffic stream and effectively set advertising rates. Their phenomenal traffic growth has glutted the ad market, forcing down rates. Digital publishers, from *The Guardian* to BuzzFeed, can stay ahead only by chasing more traffic—not loyal readers, but millions of passing eyeballs, so fleeting that advertisers naturally pay less and less for them."

A great case against free is made by *The Economist*, a magazine I have subscribed to for a decade. Prior to my London visit, I read that *The Economist* had grown its print circulation from 1 million weekly copies in 2006 to more than 1.6 million in 2015, at a time when many print publications have seen their circulation decline. *The Economist* did this while charging handsomely for both the magazine and online subscriptions, which average around $150 a year, and cost the same digitally as in print. "Our business model assumed print advertising will go away," Tom Standage, *The Economist*'s deputy editor, said over a rushed sushi lunch. "But if you give away content because you want to get ad money, when the ad money goes away, you won't be able to afford your content. We're not interested in reach. What we want is the profit!" *The Economist* was completely agnostic about print or digital, so long as readers paid for it.

Standage felt the paper edition of *The Economist* had actually grown because of something he called "finishability": the ability of readers to actually finish an issue. A magazine has a defined beginning, middle, and end, and reaching that end is incredibly satisfying. "We sell the feeling of being smarter when you get to the end," Standage said. "It's the catharsis of finishing." A news website, by contrast, can never be finished. It is a constant stream of stories, updates, and special features whose very attraction is its endless content. What Standage found interesting was that the growth of *The Economist*'s digital subscriptions has mostly been to an older audience while younger readers, like me, prefer the magazine in print. "We assume younger people want *The Economist* as a social signifier," Standage said. "You can't show others you're reading it with the digital edition. You can't leave your iPad lying around to show others how smart you are."

Standage's reference to the iPad was telling, because that device is indicative of the gap between digital publishing's great promise and reality. At first, publishing companies greeted the iPad as their financial salvation, and rushed to create elaborate apps that could unlock all sorts of interesting, exclusive content, which they figured readers would happily pay for, in a way they largely refused to with articles on websites. But these apps proved costly to develop, and more importantly, readers didn't want them. As iPad sales flattened, then fell, publishers hastily abandoned fancy tablet apps. "If you look at two or three years ago when we all believed that tablets were going to soar and some believed they would replace print, that hasn't been the case," *Martha Stewart Living* publisher Daren Mazzucca told the magazine blog *Launch Monitor* last year. "The paper format is still the primary vehicle that [readers] want to engage with. They curl up with it, take it with them, and tablets have pretty much plateaued in the marketplace."

Jeremy Leslie just shook his head when I mentioned tablets. "I've sat in enough development meetings for iPads and digital editions, where they are talking about design and user experience, that if someone walked in and said, 'I have the format! A bunch of A4 pages you can print on and flip through!' everyone would say 'Thank God! You've got the answer!' A magazine is just better."

The reasons are simple: reading on paper is highly functional and almost second nature for us. It engages those same five senses that Maria Sebregondi spoke about when explaining the appeal of a Moleskine notebook. Even though the content of an article in the print edition of *The Economist* is the exact same as one I can read on the publication's website or app, the digital experience lacks the smell of the ink, the sound of the page crinkling, the texture of the paper on my fingers. These may seem irrelevant to the way an article is consumed, but they aren't. Read on an iPad, every article looks and feels the same. The haptic variation from one printed page to another helps stem the feeling of information overload.

Digital publishers are starting to realize this, and a number have actually begun experimenting with print publications of their own. In the past few years, the music website *Pitchfork*, the US political blog *Politico*, the Jewish publication *Tablet*, and the technology culture website *Pando Daily* have all launched a variety of print products, from monthly magazines that come with limited-edition vinyl records, to tabloid newspapers distributed from boxes around Washington, DC, each week, to name several examples.

"The truth is that in this industry, paper is so much more valuable than digital," Penny Martin said, when we spoke at her office. Martin is the editor in chief of *The Gentlewoman*, a clever, biannual women's fashion magazine that is one of the largest successes of the independent British magazine industry. Each printing of more than 100,000 copies quickly sells out, and back issues trade online for many times their original price. Martin came from the world of fashion blogging, but gradually saw the digitization of journalism as a strip mining of content and ideas (a framed poster on the wall behind her read "I Blame the Internet," another has a signed *Gentlewoman* cover of Angela Lansbury). "The great hope was always that fashion ads would come online," Martin said of her time in digital publishing, "but they never did." Luxury fashion brands that advertised in *The Gentlewoman*, such as Chanel, were doubling down on print, because even the best designed online campaigns looked cheap on a screen. Print, she strongly believed, had become a luxury item. "If we're told paper is an obscene squandering of resources, then it's as luxury as leather," Martin said.

Most of the successful recent magazines in the United Kingdom have embraced a luxury approach, with production values and prices to match. The widely acknowledged pioneer of this is Tyler Brûlé, a Canadian journalist based in London, who began the glossy design magazine *Wallpaper* in 1996 after recovering from a sniper's bullet in Afghanistan. After he sold *Wallpaper* to Time Inc., Brûlé launched *Monocle* in 2007, a journal of global style, business, and other journalism. *Monocle* is a high-quality publication with substantial paper, photographs shot on analog film, and staff reporters and bureaus around the world. Each issue is more than an inch thick, and sells for $20 on average.

Andrew Tuck, *Monocle*'s editor in chief, told me that Brûlé's counterintuitive approach to publishing, at a time when larger publishers were abandoning paper for digital, quickly paid off when the recession of 2008 hit, a year after *Monocle* launched. "The crisis hit and we had the best year you could imagine," Tuck said, when we spoke at Midori House, the company's leafy headquarters in central London. Since then, *Monocle* has grown its circulation at around 7 percent annually, without devoting much effort to social media, or even giving away a single copy for free. Although *Monocle* has other businesses, ranging from lines of clothing, luggage, and books to a number of cafés, the bulk of the company's profit comes from print advertising and sales of magazines.

Tuck believes print allowed *Monocle* to build a long-term relationship with readers, because an issue of *Monocle* lives on for years after its publication. The magazine sits in the home of readers, who pick it up repeatedly and pass it along to other readers over and over again. By contrast, "if we have an article on a website or the iPad, people see it once," Tuck said. Print also delivers *Monocle* readers the value of serendipity, because the linear process of reading a paper magazine allows readers to encounter stories, images, and ideas that they would not seek out in the digital format. That sense of surprise is highly valued.

Finally, things simply look better on the printed page, especially advertisements. All of these benefits added up to a product that justified *Monocle*'s high cost to readers and advertisers. "We are very confident about print magazines in our part of the market. The pendulum has definitely swung back," Tuck said. "There is no romance in the world of digital. In a gentle way, there is a romance about the print product. It is tactile, beautiful, and you *smell* the ambition on the page. You can't smell ambition when you are on a website." Analog romance and ambition was something *Monocle* could easily sell.

———◆———

*I*f *magazines have shown a* potential vision for print's future, the path forward for newspapers is less clear (don't worry, we'll tackle books in the next chapter). The newspaper business was built on the notion of paper as a vehicle for the most up-to-date, relevant information, and print simply cannot compete with digital in this regard. If the newspaper is to remain a valuable analog medium, the very notion of a newspaper needs to be rethought.

Most daily newspapers are actively trying to decipher the right place for print in their future. One man tasked with this is Jon Hill, the chief designer for the UK's conservative broadsheet *The Daily Telegraph*, who was overseeing a huge redesign of both the *Telegraph*'s print and digital editions when we met at the paper's offices. To gain a new foothold in the postdigital publishing business, Hill also felt a print newspaper had to be perceived as a luxury product. That didn't mean adopting the glossy look and feel of a magazine, but the more esoteric personal luxury of a contained, physical reading experience, at a time when the flow of free, uncontrolled information is omnipresent.

"Reading a newspaper in 2015 should . . . no it must . . . be a pleasurable experience," Hill said, spreading open that day's *Telegraph* on a table. "It has to be something you *enjoy*. Not just the default way of getting information, but a choice." He compared a print newspaper to a library, an edifying experience that held itself as culturally superior to the noise of a Twitter feed. This rooted *The Daily Telegraph*'s editorial voice in a physical document of record, which carried a naturally superior sense of gravitas when its headlines were read in the halls of power rather than words seen on a screen. "Let's not be ashamed it's a dead tree. Let's own that."

Hill told me the *Telegraph* still made most of its money from print, which was the case for nearly every newspaper, including the *New York Times*, a publication that I pay $300 a year for digital access to (and one of the few newspapers to successfully get people to do so). In 2014, the *New York Times* conducted ethnographic research in the homes of its print subscribers, analyzing their daily or weekly ritual with the newspaper. What they discovered was that while there was a strong core of traditional print readers (such as an elderly Jewish woman on Manhattan's Upper West Side, reading her Sunday *Times* over a bagel and coffee at Zabar's), a significant number of new print subscribers were younger readers, who were actively choosing a paper newspaper.

Respondents liked the contained reading experience of the paper (its finishability), and the way stories were carefully laid out to facilitate a reader's narrative journey within a greater editorial context. On average, the research found that *Times* print readers spent more time reading stories than did digital ones. They liked the serendipity of finding, and reading, stories they never would click on with the digital version, and the ritual involved with reading certain sections at certain parts of their day. Many younger print subscribers talked about their desire to disconnect from digital devices but not the world and its information, and several mentioned keeping the paper open at the dinner table so as to spark a conversation with their kids, who were often distracted by phones and tablets.

"The appeal of print is that it's not the web," said Tom Bodkin, deputy managing editor and chief creative officer of the *New York Times*, who oversees the design of the paper and its digital properties. "Look," Bodkin said. "I love this shit. I love old technology. I collect old technology, including motorcycles and film cameras, and use it. But I don't have the illusion that it'll grow in any significant way." While print revenues were holding steady, they were definitely not an area of significant growth for the paper. What would likely occur, Bodkin said, was that print would hold on to a particular core of readers, young and old, similar to vinyl records, and those readers would form the loyal base of the brand, because they read the *New York Times* in a walled garden, on its own, undiluted by links or other distractions from competing news sources. "When you're experiencing the print *New York Times*, you're not experiencing it on a platform that delivers the rest of the world." While new digital media outlets, such as Buzz-Feed or Huffington Post, claim to have more readers than the *New York Times*, the loyalty of those audiences cannot even be compared in any serious way. One has invested in the publication's brand and identity with time and treasure. The other clicks in for a glance at a tantalizing headline. BuzzFeed may have hired some excellent writers and occasionally produces some real quality journalism, but funny lists and articles about taste testing nine different brands of cat food (I swear) still drive the bulk of its traffic.

Digital subscriptions like mine gave the *New York Times* enough financial stability to let the print edition settle into this new, smaller niche. But other newspapers who gave away their news online were now stuck between a shrinking but cash-generating print past and a growing digital future that bled money. No paper has faced more of a challenge with this paradox than *The Guardian*, which has vocally staked a position as the leader of digital newsgathering. "There is no confusion on where this organization is traveling," said Robert Yates, the associate editor of *The Observer*, the Sunday paper of *The Guardian*, speaking over a pint in the pub downstairs from the company's swank new headquarters by King's Cross station.

"Holding on to print for us is not necessarily an aesthetic choice. Good old-fashioned print is where most of the revenue comes from . . . the projections of online revenue were optimistic to say the least." *The Guardian*'s print readers read two and a half times more than digital readers do, and were still worth more to its advertisers. But their numbers were shrinking, and like many of his peers in London, Yates also felt that if the print newspaper was to have a future, it would be as a luxury product. "When Sunday paper magazines kicked off in the 1960s, they were luxury products, driven by smart ad agencies for a new aspirational audience, who needed an outlet for that new creative content."

The Guardian had been experimenting with the marriage of new technology and business models to try to create a print newspaper suited to the future. In 2013 *The Guardian* launched *The Long Good Read*, which was a printed digest of the previous week's most read, commented on, and interesting stories to appear on *The Guardian* website, automatically chosen by algorithm and distributed free at *The Guardian*'s east London coffee shop. *The Long Good Read* eventually morphed into *Contributoria*, an ambitious crowdfunded online journalism platform, where editors proposed topics, freelance journalists pitched ideas, and readers directly funded the stories they wanted written. Each month, the best stories from *Contributoria* were printed out in a newspaper, sent to subscribers of the service, and packaged with select editions of *The Guardian*.

"With digital, you get a massive audience with global reach, but the experience of reading and digesting that you can get out of print experience is still unparalleled," said Matt McAlister, *The Guardian*'s general manager of new digital business and its CEO of *Contributoria*.

"I think a lot of people who really value the reading experience are missing print and want print." McAlister is far from an ink-stained wretch of the British newspaper old school. He is American, and his background is with Silicon Valley media companies, including Yahoo! In many ways, McAlister is the man most responsible for *The Guardian*'s headlong push into 1's and 0's, and his solution to print's purpose is based on the premise of newspapers that are printed on demand. "With projects like *Contributoria* or *The Long Good Read* it's very easy to imagine a world where [all] newspapers are operating like these services," McAlister said, eight months before he pulled the plug on *Contributoria*.

The company that made both *The Long Good Read* and *Contributoria* experiments possible and is driving this on-demand future is Newspaper Club, based in Glasgow, Scotland. Newspaper Club was created out of necessity in late 2008, when a trio of friends (Tom Taylor, Russell Davies, and Ben Terrett) made a newspaper called *Things Our Friends Put on the Internet in 2008*, which they created as a gift for the fifty guests who were attending their office Christmas party. True to its name, the paper was filled with stories, images, and other thoughts their guests had posted online. "We were interested in physical stuff," said Taylor, a software designer who now works for the business card company MOO. "People who were working on the web, like us, weren't hung up on the digital thing anymore."

When the trio began calling British newspaper printers to get quotes on printing *Things Our Friends Put on the Internet in 2008*, they ran into a problem. Newspaper printing presses were set up to print large quantities of newspapers at once—tens of thousands of copies—and even though these printing plants had been greatly diminished by a decline in newspaper sales, small print runs were uneconomical. They finally found a printer willing to print one thousand copies of *Things Our Friends Put on the Internet in 2008*, gave a bunch away, and put the rest for sale on their own blogs. Surprisingly, they sold out.

The three friends had stumbled on two crucial things that great businesses are made of: a market for a product (custom newspapers), and underused capacity in the industry serving them. If they could provide an easy, turnkey way to design a newspaper and get it printed, then newspapers could be accessible to anyone. The key to this was the advent of digital newspaper printing, which was a relatively new technology that used a giant inkjet printer to print on newsprint, and allowed newspapers to be printed in much smaller quantities. By standardizing layouts in a certain format, it was possible to economically print even a single copy of a newspaper. "That's very interesting," Taylor said, "because that makes new types of product completely possible."

Their initial business was a service called Paper Later, which allowed users to save articles from the Internet and print their own custom newspaper. But Paper Later quickly gave birth to Newspaper Club, which is a broader turnkey newspaper-publishing solution. Anyone can upload a design to Newspaper Club's website, select the size, style, and quantity of a newspaper, and have it automatically printed, bundled, and shipped anywhere in the world. The newspapers they print range greatly, from student art projects, high school newspapers, arts and culture journals, and hyperlocal newspapers to catalogs for clothing brands, business presentations, programs for music, art, or culinary festivals, and even wedding souvenirs. On a given day, Newspaper Club could be printing newspapers designed for a major retailer by an ad agency, or by a teenager in their bedroom. In many ways, Newspaper Club is the analog equivalent of a blogging platform. It opened newspaper printing from a closed product to something anyone can create. The company's motto is "Print's Not Dead."

Newspaper Club is now run by Anne Ward, a former librarian and soft-spoken travel book writer from Glasgow. The company had printed as many as seventy thousand copies of a single paper, but most of its projects still involve ten copies or less. When we met for lunch in London, Ward told me that Newspaper Club was about to print off its eight-millionth newspaper. "Those eight million are 1's and 2's and 5's," Ward said. "We celebrate every million paper, and those celebrations are getting much closer together." The company had been growing by over 40 percent a year, and had been profitable since 2013. Their customer base was branching out from the United Kingdom and growing quickly in North America. "We're not involved with news at all. [Traditional] newspapers are a terrible business. They print a lot of rubbish, and people have lost a lot of faith in them," Ward said, referring to the tabloid hacking scandals that had recently gripped Britain. "But newsprint is something people are attracted to. It is a cheap, effective way of publishing, and as the printer's work has declined, the opportunity to do small jobs has increased. That's our strength."

Eighty percent of Newspaper Club orders are for papers under three hundred copies, and these are fulfilled by a digital printer near London. For anything larger, the newspapers are sent to Sharman's, a traditional newspaper printer located an hour north of the city, in the town of Peterborough. The day before I met Ward, I caught a train up there to speak with Mark Sharman, the fourth generation of his family running the business, which began back in 1910. Mark Sharman is in his early forties, and has the shaggy hair and wide smile of Charlie Watts, the drummer from the Rolling Stones. His great-grandfather had built the business printing large community newspapers for towns and counties, typically with print runs of fifty to a hundred thousand copies at a time. In the early 2000s, newspaper conglomerates were consolidating globally, scooping up local papers and merging their operations. As a result, Sharman's lost its largest contract in 2006, and 60 percent of the printer's business disappeared overnight. Mark Sharman walked me over to the wall of his office, where a map of the United Kingdom was littered with small dots representing other newspaper-printing plants in operation. Since he joined the business in 2002, the number of dots on his map had declined by more than half.

To survive, Sharman's accepted smaller jobs that other printers refused: school newspapers, local council newspapers, brochures, and bulletins. A few years back, he attended a music festival and discovered Newspaper Club, which had printed the concert program. He found their untraditional approach interesting, and struck up a deal for Sharman's to handle Newspaper Club's larger orders. Initially that involved one or two print runs a week for Sharman's, but now it prints as many as twenty-five Newspaper Club publications a week, representing 20 percent of Sharman's total print runs and a much larger percentage of its profits. Last year, Sharman's bought a controlling stake in

Newspaper Club, chiefly to protect its own business with them, which is fast becoming the company's main area of growth.

On the floor of his plant, a giant newspaper press rapidly unspooled huge rolls of paper, which sped and twisted over and around, rubbing against inked printing plates, to produce the latest Newspaper Club newspaper: a full-color program for an upcoming food festival in Ireland. The first folded newspapers emerged from the end of the press a minute later, and Mark Sharman plucked one copy from the line every thirty seconds, quickly flipping through the pages to check color, contrast, and resolution, as his staff made adjustments. "Without Newspaper Club, it would be very difficult for us," Sharman shouted over the drone of the press, while thousands of newspapers piled up behind us. "We may be in newspapers, which are declining, but we can be in the growing part of the business."

As the printing press churned out newspapers, I looked around the Sharman's warehouse at small piles, large bundles, and heaving pallets of other Newspaper Club newspapers, from a graphic design student's thesis to *The Bedford Clanger*, an arts and culture paper for the small city of Bedford that Erica Roffe created in 2011 with just forty copies, after discovering Newspaper Club online. She now prints twenty-five thousand copies a month. "I think that the ease with which you can print through the Newspaper Club has seen an increase in grassroots publications," Roffe wrote me in an e-mail. "While national newspapers are struggling, more niche publications are thriving."

Looking onto this unending sea of newsprint, I thought back to the question "Why a book?" that had brought me to London. There were strong logical reasons that people chose print—readers paid more attention to it, advertising worked better on it, it looked nice and its financial model was straightforward—but what struck me was how irrational my own answer to the question was.

I wrote books and articles for print publications out of a love of the printed page. It was a love that was introduced by my parents, and one that they nurtured my whole life as they read me books and bought me magazine subscriptions, filling our house with every possible type of printed publication. Seeing that same love of print capture my own daughter each night when she runs to her bookshelf to pick a bedtime story restores my faith in humanity. As the twelfth-century Judaic scholar Judah ibn Tibbon said nearly a thousand years ago, "Make books your treasures and bookshelves your gardens of delight."

A love for the printed page brought me into this cockamamie analog industry. I write to fill the pages others will read. That is what I do. I am a reader and a writer. But to write without having that final transfer of ink to paper, to publish without actually printing off a physical publication, and then having someone pay to bring those pages home with them, is devoid of any true sense of accomplishment. Why a book? Why print? Because it is real. Because I can hold this very book in my hands once it is printed, see my name on the cover, and know that all the work was worth it, regardless of how many copies it sells. That feeling is the ultimate luxury, and it is one I will happily pay for, time and again, as a reader and a writer.

"You can't beat a physical thing you can pick up, leave somewhere, wave, and read on the Tube," Ward told me, as she picked up a Newspaper Club–printed paper at an art exhibit we were exploring in London. "We've definitely made it possible for people to turn that idea into a reality."

6

The Revenge of Retail

The men from UPS and FedEx wheeled stacks of cardboard boxes into Book Culture's storefront on Columbus Avenue in a steady stream, while the staff did their best to shelve the contents before the towers of boxes scattered all over the floor grew even taller. Even though Book Culture would not officially open for another three days, it was already full of curious customers from the neighborhood. Every fifteen minutes or so, someone would walk in from the rain, step through the doors, and blurt out their astonishment at the very sight of the place.

"Oh, a bookstore!" said a woman wearing a poncho, with tre-mendous surprise.

"Isn't it wonderful that there's a bookstore here now?" asked an-other lady, to no one in particular.

"This is great," a man said, frozen in the doorway.

"This makes me so happy!" said his friend, equally astonished.

Their amazement was spontaneous and not the least bit con-trived. The Upper West Side's bookstores were supposed to be extinct. If they were mentioned at all, it was when they closed. To enter a brand-new bookstore in 2014, in New York City of all places, seemed as improbable as spotting a woolly mammoth from the nearby American Museum of Natural History grazing in Central Park. And yet here they were, stepping into Book Culture's 3,000-square-foot location (its third), watching other people buying real books. This wasn't just a bookstore opening up. It was a symbol of hope, a lone flower poking up from the spring frost after a long, brutal winter for bookstores.

Along with record stores, bookstores were among the first brick-and-mortar retail segments challenged by digital technology, starting when Jeff Bezos launched his online bookstore Amazon out of a Seattle garage in 1995. Amazon grew to define the limitless power and speed of online retailing, devouring books, then other products, until it became the Internet's largest retailer. Amazon began an e-commerce revolution that seemed to upend the world of retail in every sector, from eBay and auctions, to Craigslist and classified ads, to Fresh Direct and groceries. An online retailer could deliver seemingly anything right to your door, faster and cheaper than a physical store, with free shipping included.

"Retail guys are going to go out of business and e-commerce will become the place everyone buys. You are not going to have a choice," the venture capitalist Marc Andreessen said in an interview with the technology website Pando. In Andreessen's future, no one would shop at brick-and-mortar stores, and the numbers were his proof. In 2000, US e-commerce picked up $50 billion in sales. By 2014, that number was $350 billion. "We're still pre-death of retail, and we're already seeing a huge wave of growth. The best in class are going to get better and better," Andreessen predicted. "Retail chains are a fundamentally implausible economic structure if there's a viable alternative."

Bookstores seemed to bear the brunt of this, as Amazon captured nearly one quarter of the book retail market (worth, some estimate, more than $5 billion) with cutthroat pricing, unparalleled selection, and technological innovations ranging from user-generated book reviews and ratings to software algorithms that can recommend titles, and of course the eBook market leader Kindle, with its wireless access, limitless capacity, and one-tap purchasing.

Thousands of bookstores have closed over the past twenty years in the United States alone, with similar declines in other Western markets. These stores ranged from small independent bookstores in cities and towns to large publicly traded chains with hundreds of locations, including Waldenbooks, B. Dalton, and Borders, which had over six hundred stores at its peak. Barnes and Noble, the last of the big bookstore chains in the United States, has reduced the number of its stores around the country by 20 percent. *Bookstore* is now a word that most often appears next to *decline, death, end*, and *doomed*.

But a funny thing happened on the way to the funeral. Bookstores began, against all expectations, to grow once again. Sales grew steadily from a low point in the recession, but more important, so, too, did the number of bookstores. This has been especially true of independent bookstores, those beloved mom-and-pop operations romanticized in such movies as *You've Got Mail*. From a low point in 2009, of 1,650 stores (down from a height of 4,000 in the 1990s), the American Booksellers Association (ABA) noted a marked increase in new member stores, which had steadily grown to 2,227 in 2014, with no sign of slowing down. Although the ABA counts a significant number of the nation's independent bookstores as its members, it is just a fraction of the whole market, and various estimates put the total number of US bookstores closer to 13,000, up from a low of around 10,000 in the early 2000s, when the last US census counted them.

Book Culture's new Upper West Side location was just the latest bookstore to open in New York City in recent years, and one of the few in Manhattan to do so. If it could succeed in the most expensive, demanding, and unforgiving retail market in the world, it would not only be a symbolic victory for the bookstore's resilience against Amazon, but a telling lesson of the fundamental advantages that brick-and-mortar retail still held over e-commerce, whose leading brands, such as Apple, Warby Parker, and even Amazon, were coming to see that building a profitable retail business without physical stores was nearly impossible.

———•———

"*The world isn't so much* a competitive war of distribution between brick-and-mortar stores and online retailers, but a war of brands," Bob Grayson said, as we walked from his company's office, located next to one of the last big Barnes and Nobles in New York on Union Square, to check out some of the retailers nearby. "Brands that can get in front of consumers in a multitude of ways have a tendency to overpower." Grayson has been a retail man for more than forty years, and his firm, the Grayson Company, consults with such brands as Victoria's Secret, Abercrombie and Fitch, Tommy Hilfiger, Reebok, Citizen, Lululemon, and even Etsy. Grayson is a firm believer in the theory of omnichannel retail, where you sell your products in as many places as possible, from stores and e-commerce to Avon-style peer marketing, discount outlets, and even print catalogs. In all, Grayson counted twenty-one different possible channels where brands can sell to customers.

Of these, e-commerce has received the most attention, due to its rapid growth over the past twenty years to more than 7 percent of US retail purchases in 2015, according to the US Department of Commerce. In a tight retail market, that is truly rapid growth, and I can attest to its prominence. These words appear on a monitor I bought from dell.com, which sits on a desk I found on Craigslist. I booked my flight to New York on hipmunk.com, and took notes in Ecojot journals I ordered online with a pen sent to my door by staples.com. Each year, my credit card statement shows well over a thousand dollars' worth of purchases from Amazon, including books, eBooks, and other items, and I am a comparatively light online shopper. Some friends of mine almost never set foot inside a store. When he lived in New York, my brother once bragged about ordering a single battery from Amazon's one-hour delivery service, even though he could have walked to the nearest store, bought one, and returned home in five minutes.

"Purchases that are based on convenience and price are two competitive edges that have driven the growth of online retailing," Grayson said, as we entered a Lululemon store on Union Square's west side. The web puts together customers and retailers who'd have trouble finding each other, and allows them to price shop instantly and globally. "Those transactions have been extracted from brick-and-mortar retail. The web pulled them out."

What the brick-and-mortar retail store does best is deliver an experience, something online retailing struggles with, which is why Grayson was adamant that physical retail remained the central hub of an omnichannel retail strategy. We walked around the store, and Grayson pointed out various things Lululemon had done to create a shopping experience: the spa music, a chalkboard listing yoga classes, the greeting by the young salesperson wearing the company's fitted clothing, a filtered water fountain, and bowl of dog treats. "You get so many touch points in a store that you just can't communicate on a website," Grayson said.

Outside, we walked through the Union Square Greenmarket, one of the country's pioneering and busiest urban farmers' markets. It was situated across the street from a massive Whole Foods Market with infinitely greater selection, prices, and the convenience of a controlled environment, in a city where dozens of online food delivery services can bring the world's freshest, most varied ingredients and meals to your door, but the Greenmarket was packed nonetheless. Grayson felt that the same factors that fueled the growth of farmers' markets in the United States (up from 2,000 in 1995 to more than 8,000 in 2015) were now driving the revenge of bookstores, including the Strand, a block away. With its towering shelves, creaking wooden floors, and quirky staff spread over three floors, the Strand is perhaps the most romantic and famous bookstore in New York. But that hasn't made it immune to the numerous challenges independent bookstores have faced over the past three decades.

"I look back from the time when I took my first publishing job in 1982, and every single decade there's been some reason for people to proclaim that independent bookstores were gonna die," said Donna Paz Kaufman, a former bookseller, whose company, Paz and Associates, consults with new and existing independent bookstore owners nationwide. The first challenge was mall chains, such as Crown, Encore, and Tower, which provided a bigger selection and lower prices than independents could match. These were quickly surpassed by the big-box chains—Barnes and Noble and Borders in the United States, Indigo and Chapters in Canada, and Waterstones and WHSmith in the United Kingdom—which erected warehouses filled with books that they sold at the lowest price possible, extracting the best terms from publishers beholden to their market share. Long before Jeff Bezos registered Amazon's domain name, the big-box chains, as well as other discount retailers such as Costco and Walmart, greatly reduced the ranks of independent bookstores. Amazon was the final blow, not just for smaller independent bookstores, but more dramatically for the big-box chains, which could not compete in the two areas (price and selection) where they once held an unassailable advantage.

"The Strand had faced artillery coming in from all directions," Grayson said, as we stepped into the store's narrow elevator. By 2012, the Strand had weathered this storm (largely because it owned the building), but the Bass family, which founded the Strand in 1927 and still operates it today, turned to Grayson for help. His approach to the Strand was to emphasize, rather than water down, the venerable bookstore's analog charm. Grayson suggested changes to the store's layout and merchandising, including more books displayed on tables to encourage easier browsing, a new vinyl record section, and a home library setup service, with books chosen for a client's literary interest or by color scheme (all-white libraries were hot in the Hamptons).

Other new bookstores that have opened up in the past few years in New York and elsewhere still approach bookselling in a way that hasn't markedly changed since Gutenberg invented the printing press. The big difference is that they are able to turn perceived analog weakness into strength, and sell that as a desirable lifestyle choice to customers. This new generation of bookstores all define themselves as an enlightened, more pleasurable retail alternative to Amazon and big-box stores. They are warm, inviting, often beautiful spaces, with friendly, knowledgeable staff, refined inventory, and a sense of place. Most support local authors, hosting reading groups, book clubs, and nightly events. They take all the things Amazon sees as liabilities (physical real estate, human workers, limited inventory) and turn them into assets.

"Retail used to be 'I need something, I need to go get it,'" said Donna Paz Kaufman, who noted that her company is working with more new bookstores, and teaching more courses to prospective bookstore owners, than at any other time in the previous twenty-four years. "Today retail is going out to browse. Retail is more a sense of place and an experience than a destination for commodities. That shift has been really profound. For the independent bookstore world to dance with that change is fairly easy," she said.

*B*ookstores *that had recently opened* around New York provided hints about the forces behind the larger revenge of retail. The acknowledged pioneer was McNally Jackson, which Sarah McNally (whose parents own the popular Winnipeg, Manitoba, store McNally Robinson) opened in 2004 in SoHo. Located two blocks from one of the busiest retail strips of Broadway, McNally Jackson set itself apart from other bookstores. It featured a zigzag of custom wood shelves, a large café, soft lighting, cool music, and a vibe that made reading seem genuinely sexy.

In 2009, the woman in charge of events at McNally Jackson, Jessica Bagnulo, opened a small bookstore, called Greenlight, in the Fort Greene neighborhood of Brooklyn with her business partner Rebecca Fitting (who worked in publishing). At the time they couldn't get any financing from the bank, so they sold debt in the store to members of the community (the loans were paid back, with interest). Bagnulo and Fitting both knew the challenges of selling books in the age of Amazon. But they also saw how bookstores held distinct advantages over digital commerce.

For one, they can sell.

"We have a hand-selling culture here," Fitting said when I visited Greenlight, a warm, intimate store with a stunning central wraparound bookcase that anchors the space like a great ship docked with literary cargo. *Hand selling* is a book industry term, which essentially means that sales associates will place books people want to read into their hands. This involves basic human skills such as reading body language, making eye contact, inquiring about personal taste, and using human judgment to suggest the right book. Amazon doesn't do hand selling; its software algorithms recommend titles with calculations that weigh the likeliness of what you would like to read, based on what you read before and what others who read those books also bought. Most often I find Amazon just feeds you predictably similar titles. Big-box bookstores often fail at hand selling, too, because their selection is too vast and the staff inexperienced. "Hand selling is one of the things that independent bookstores do best," Bagnulo said. "That's when you put the book in someone's hand and say, 'I love this, and I think you'll love it.' It's not about pushing 'the right literature,' it's about having that conversation."

Bagnulo described her ideal employees as book lovers who are good with people (not always the easiest combo to find). They are trained to walk a fine line, so that a hand sale feels like a natural end to a conversation, rather than a pushy transaction. Two of the best books I bought last year were hand sold to me by Jess Pane, a book-seller at Greenlight. I told her what I wanted (a new author who was clever, and something for my daughter, then one and a half) and she placed two titles into my hands: *The Dead Fish Museum*, a collection of darkly funny short stories from Charles D'Ambrosio, and *Bronto-rina*, a picture book about a brontosaurus that dreams of being a balle-rina, which remains a bedtime staple.

Elizabeth Bogner, the manager at Book Culture's new store, told me that hand selling a book was a unique interaction in the world of commerce. The week before, she had recommended Elie Wiesel's brief but powerful Holocaust memoir *Night* to a customer who wanted something historical but readable. He returned the next day, waited nearly an hour for Bogner to arrive, told her he devoured *Night*, and wanted another recommendation. She handed him two options, sat him in a chair, and told him to read five pages of each. He left with *Moby Dick*. "Amazon wouldn't have recommended that," Bogner said. It would have spit out a dozen more Holocaust memoirs. "You engage with someone by finding out how someone else's words made them *feel*! That's ultimately not translatable to an algorithm."

As much as we believe that limitless selection is desirable, we actually crave limits as shoppers. According to scientists such as Barry Schwartz, author of *The Paradox of Choice*, we are paralyzed and even terrorized by endless options, which is exactly what Am-azon gives us. Choosing from every book ever published seems like a dream, until you're forced to sift through hundreds of thousands of titles on your Kindle, and all the reviews attached to them, hoping to find something good.

"It's the distinction between browsing and searching," said Adam Alter, an associate professor of marketing at NYU Stern School of Business, who studies consumer behavior. I met Alter at a party in New York, and told him about a book I had stumbled upon at Book Culture earlier that day. It was called *The Lonely Typewriter*, and I only found it because it was displayed on a shelf next to where I had placed my coat. The book is the story of a typewriter that is forgotten by a woman, only to be rediscovered years later when her son's computer breaks in the middle of a homework assignment. Basically it is *The Revenge of Analog* for kids. I took one look at it and immediately brought *The Lonely Typewriter* to the cash register. Alter laughed. "There isn't room for serendipity online," he said. "The strongest reward structure for humans is variable reinforcement, like gambling. That's the most addictive kind of reward signal, when you can't predict it." An impulse buy delivers a jolt of instant gratification coupled with the pleasure of a surprise discovery.

The second advantage brick-and-mortar retail stores have over e-commerce is merchandising, the myriad ways to highlight what's for sale. These range from shelving and racks to lighting, music, decor, and even scent. All these matter, because shopping is a full-body contact sport. "If you still don't believe all this," wrote the retail consultant Paco Underhill in his classic book on retail psychology, *Why We Buy: The Science of Shopping*, "go to the home of a product fairly unconcerned with smell, touch, or any other sensual experience—a bookstore. There you'll be greeted to the sight of shoppers stroking, rubbing, hefting and otherwise experiencing the physical nature of a product where no physical attribute (aside maybe from typeface size) has anything to do with enjoyment. Still, helplessly, we touch. We are beasts like any other, and despite all our powers of imagination and conceptualization and intellectualization and cerebration and visualization, we physical creatures experience the world only via our five senses."

Bookstores merchandise their products with comfortable chairs and private reading nooks, elaborate window displays, curated sections of books (Greenlight has one for local independent publishers, for example), beautiful shelving, clever lighting, and random touches of whimsy that set the mood. When I asked why Book Culture's co-owner Chris Doeblin was hanging a huge model biplane above the stairs leading to the kids' section, he turned from the ladder and said, "Because you've just got to have a Sopwith Camel!" with the conviction of Snoopy. Book Culture even has a writing desk, with free stationery and stamps, for no other reason than to get customers to sit down and write a letter they will remember sending from the store.

E-commerce designers use the term *rich browsing experience* to describe the quality of a website or app's design, but even the most basic corner store is a vastly richer browsing experience than anything you can view on the web. "If you follow most retail online, it's pretty much the same exact format," said retail designer Daniel Gonzales. Online merchandising is basically a digital version of the most mundane catalog: grids or lists of products, with images, prices, brief descriptions, and reviews. There's little room for creativity or uniqueness, because the goal is maximum efficiency.

One of the clients Gonzales has designed stores for is the California clothing brand Alternative Apparel. I had never heard of the company before, but I happened to walk by its SoHo storefront while in New York, and a red hooded sweatshirt in the window caught my eye. Even through the glass, I could tell this was a luxuriously soft, incredibly comfortable sweatshirt. I went inside, asked the salesman about the different sweatshirts sold there, tried on two, and left three minutes later with the red one in the window, which is the best article of clothing I have ever owned (actually, I'm wearing it as I write this). I have purchased two more Alternative Apparel sweatshirts since then, and sent several friends to the store, all because I looked up from the street. I was sold before I even walked in. That's the "tactile advantage of physical shopping," according to *Retail* magazine's design editor, Allison Medina. "You can't touch a dress on an iPad, or smell a cantaloupe through your computer to see if it's ripe."

---

*T*he *assumption behind e-commerce is* that none of this really matters in the end. Consumers are savvy, and the laws of market capitalism will always reward those delivering the best price and selection, regardless of such frills as beautiful spaces, salespeople, and merchandising. As consumers use their computers more often to shop, the vestiges of that archaic culture of shopping . . . physical stores . . . will fade away as the advantages of digital retailing in selection, pricing, and infrastructure triumph.

That assumption is false.

While selling something online is relatively easy, making a profit from e-commerce has proven incredibly difficult. Name an online retailer, and odds are that company is losing money, regardless of how many billions in sales they pull in. Amazon only recently turned a profit on its retail division after twenty years of selling books and other goods online, and it was hardly a robust one. In the second quarter of 2015, Amazon's retail operating margin in North America was 2.5 percent, compared to 25 percent on its more lucrative web services. Groupon continues to spend more than it takes in on flash sales. Fashion site Fab.com blew through $336 million in funding and was sold for just $15 million. Gilt Groupe, valued at $1 billion in 2011, has been selling off assets to try to scrape by. The list goes on and on. Despite all their supposed advantages of scale, revenues, and cost (no salespeople, no rent, no inventory), why do e-commerce companies, which were supposed to replace retail stores, fail at the one thing they're supposed to do?

Let's start at the beginning: customer acquisition. A physical store can attract customers with a vast arsenal of advertisements (online, billboards, newspapers, leaflets, flyers), marketing materials (catalogs, signs, contests, sales, promotions, loyalty cards), and merchandising tactics (window displays, store design, location). An e-commerce retailer can really only reach customers through online marketing, whether e-mail blasts, search engine optimization (SEO), or digital ads. These are great at getting a message to a huge number of eyeballs, but they compete with all the other ads, content, and images vying for your attention on a small screen, and are easily dismissed or blocked by software. Converting that into a purchase is incredibly difficult and costly. Alternative Apparel could have hit me with all the digital marketing in its arsenal, but none of it would have been as effective as putting a sweatshirt in a window. Analog customer acquisition is vastly more cost-effective.

Next is the shopping experience itself. The goal of most online retailers is to get customers in and out with as few clicks as possible. The more time you spend online shopping, the more stressful it gets. Options quickly multiply, comparisons are endless, and you get sucked into a wormhole of reviews, counterreviews, and ratings. Analysis paralysis sets in. There are few things I hate more than sifting through hundreds of Trip Advisor reviews as I try to read between the lines to determine whether the hotel I'm thinking about booking is a paradise or hellhole. Others likely feel the same way. In 2014, 18 percent of American travelers used traditional travel agents, compared to 12 percent in 2013.

What these shoppers are seeking is other people to assist them, and digital shopping, by definition, is devoid of human help. Instead, e-commerce purposefully offloads the burden of shopping onto the consumer. It makes us sift through information, compare styles and prices, and then write reviews for the products. The e-commerce business may see salespeople and other humans as obstacles to be overcome, but people exist to smooth the shopping experience and make it more profitable, not less. A few years ago, after a decade of widespread implementation across the supermarket industry, several big grocery chains, such as Albertsons and Kroger, actually replaced expensive automated checkout kiosks with human cashiers, because they found the humans worked faster, customers preferred them, and theft actually decreased.

In-store shoppers spend more time and money shopping, and return to buy more often, than online shoppers. This holds true for stand-alone e-commerce sites, such as Amazon, as it does for omni-channel retail sites, such as gap.com. "Anybody who's done any research whatsoever, whatever the lifetime value of a customer is of a single channel, if you can get them to shop multiple channels, they'll be three to five times more valuable," said Glen Senk, the former CEO of Urban Outfitters, who currently runs the retail investment firm Front Row Partners. Of all the channels available to retailers, Senk said, "the weakest value is e-commerce."

Finally, there is the question of product delivery.

If you bought this book on Amazon, it was dispatched from a warehouse to a distribution center by a postal service or a courier, which sent it around to various warehouses, on various trucks and airplanes, until it arrived at your door. Not only is the gratification from the purchase delayed by hours, days, or even weeks (depending on delivery times), but if it was the wrong book, you have to now figure out how to return that package to a company without a store. Most online retailers talk about how great and easy this is, but any-time I've had to return something to an online vendor, it is a head-ache. Printing packing slips, taping up boxes, and schlepping to the nearest UPS store is a multistep process that is the opposite of con-venience. Returns tend to happen more with online purchases than in-store ones, because customers have never actually held the mer-chandise, such as the sleek leather oxfords I ordered from Zappos for my wedding, which emerged from their box looking perfectly suited to a cheap pimp. In contrast, when a pair of boots I'd bought at a store near my house began splitting at the sole a year after I began wearing them, I took them back to the store and showed them to the manager. She gave me a new pair on the spot, even though it was against the store's return policy.

Shipping is e-commerce's Achilles heel. In a retail store, the cus-tomer bears the cost of delivery. If you purchased this book at a book-store such as Book Culture, you picked it up, paid for it at the cash register, and transported the book home. If you bought it on Amazon, it was likely shipped for "free." Except there is no such thing as "free shipping." The UPS delivery driver and his colleagues don't work for free, his truck isn't free, and it doesn't run on free gas. All these things cost money, and the online retailer that offers free shipping (nearly all of them do) is really offering to eat the cost of shipping. This is the most significant reason that online retailers such as Amazon strug-gle to profit, despite huge revenues. Amazon's net shipping costs have increased by more than 25 percent a year since it launched Prime in 2005 (a premium service that includes free shipping), and those costs continue increasing exponentially. During the Christmas season of 2014, two thirds of Amazon packages were shipped for free. That number used to be one third. Again, remember that "free" basically means shipping paid for by Amazon's shareholders.

According to Scott Galloway, a professor of brand strategy and digital marketing at NYU, Amazon is playing a game of Last Man Standing. Galloway, who also has experience running online retailers, made his point in jargon-y terms. "I would argue that the total universe of profitable pure-play e-commerce retailers over the medium term have been zero," Galloway told me. In other words, simply being an e-commerce site doesn't make money. Amazon is the best example why. The company sells goods, such as books, at less than cost, then ships them for free, all the time bleeding money on shopping in the hopes that it can gain enough market share to kill off all its competitors. What's the endgame here?

"Business is about making money," said Joel Bines, a retail turnaround consultant with the firm AlixPartners who has worked to bring failing or insolvent retailers, such as Pier 1 Imports, back into the black. Bines has looked at the balance sheets of hundreds of retailers and what he sees, consistently, is that online retailing acts as a drag on the bottom line. "Being everywhere lowers margins, gives consumers all the power, drives up your costs, makes them price driven customers, and kills their affinity to the brand. All of that is being driven by online," Bines said. "Eventually someone's going to have to make money."

Last June, the *New York Times* reported that research by the firm eMarketer showed a decline in the rate of domestic e-commerce growth, which fell a bit more than 2 percentage points from 2013 to 2015. The same report predicted the growth of online shopping would slow consistently going forward. "By that point, e-commerce in the United States will still be less than 10 percent of all retail," the article said. "The death of the physical store, so widely anticipated in the last few years, is nowhere in sight."

Galloway already sees the solution for Amazon and other so-called pure-play online retailers struggling to turn a profit, and it has nothing to do with software, drone delivery, or virtual reality. It is stores. "Revenge of the brick," Galloway said. "Retail is the new black." The reasons are obvious, and pioneered by traditional retailers with e-commerce divisions. E-commerce statistics obscure an important fact: half of online retail sales in the United States are actually tied to physical stores, from macys.com and homedepot.com to June Records' putting rare vinyl records up for sale on the online marketplace Discogs, which processes nearly $100 million record sales annually. Macy's, Nordstrom, Best Buy, Apple, and thousands more have discovered that store locations are the best assets for online retail. They serve as a network of regional warehouses for online orders, reduce shipping costs, and act as a convenient pickup and return location for online shoppers. "That's more convenient than ordering from Amazon and finding a yellow sticky from UPS on the apartment door that said you missed your delivery," Galloway said.

The world of e-commerce is reluctantly coming around to the idea that it needs to open stores to survive. Some online retailers who have created brick-and-mortar outlets include the men's clothing brands Bonobos and Frank & Oak, the cosmetics company Birchbox, jeweler Blue Nile, Google, Alibaba, and a handful of others. Last November, even Amazon entered the real world by opening a 5,500-square-foot bookstore in Seattle. Although it featured some innovative displays and selections that reflected online data, Amazon's first physical bookstore was ultimately just a really nice brick-and-mortar bookstore.

"We went far and fast off the deep end on digital, and it felt the whiplash," said Eurie Kim, a principal with the San Francisco venture capital firm Forerunner Ventures, one of the largest backers of e-commerce brands, including Bonobos, Serena and Lily, Birchbox, Dollar Shave Club, and more. "Now it's coming back." What changed, Kim felt, was the desire of consumers. At first, they wanted total online access and convenience, but over time, the loss of the immersive shopping experience became apparent, especially for brands that only existed online. "Consumers say, 'I want community feedback and excitement!' Well, you can't do that online," Kim said. "You do that in a store."

On the surface that doesn't appear as cost-effective a customer acquisition strategy as targeted Facebook ads, but over the long run, the foot traffic, visibility, repeat customers, and good old-fashioned word of mouth a store creates for a digital brand have not only led to in-store growth for these brands, but greater online sales as well. Many in the technology industry, including other venture capitalists who invested in these companies, dismiss the stores e-commerce brands have opened as irrelevant marketing stunts. The real money will still be made online, they insist. Kim just shook her head at this. "They work," she told me of these new offline stores. "They sell." A lot.

The most successful Forerunner-backed case for this is Warby Parker, which is also the online retailer most aggressively pursuing a brick-and-mortar strategy. The company began in 2010, designing and selling cool glasses online for less than $100. Warby Parker shipped consumers up to three pairs of glasses for free, which they could return at no cost. In 2013, after succeeding with a series of temporary pop-up shops, Warby Parker opened a 2,000-square-foot retail store in New York. It featured two huge mirrored walls of glasses, which customers were encouraged to grab, try on, and model in a photo booth. "Originally they thought it was going to be a brand play, a kind of loss leader," said Anthony Sperduti, who works with the branding firm Partners & Spade, which helped design Warby Parker's stores. "But it did so incredibly well, they shifted to more of a retail strategy."

There are now around twenty Warby Parker stores around the United States, and the company is aggressively moving to build more. Each is designed with an eye to an immersive, memorable, and largely analog retail interaction. Young, helpful sales clerks wear matching blue coats; there are shelves of books for sale, free pencils and erasers, reading maps of nearby neighborhoods, and towering walls of glasses with mirrors. Warby Parker stores now sit behind only Apple and Tiffany in retail sales per square foot, according to various sources, which is astounding, considering Warby Parker exclusively sells $100 glasses, compared with $1,300 laptops and $25,000 engagement rings.

For all the growth of online commerce, the retail store remains more powerful than ever. The technology industry should know this, because its leading company, Apple, is the poster child for brick-and-mortar retail. Steve Jobs launched the Apple Store in 2001. Analysts called it a desperate move for the company, and one predicted the stores would be out of business in two years. Instead, the stores quickly became profitable in that time, and the Apple Store grew into the single most successful retail store on earth, grossing nearly $5,000 per square foot annually, with each Apple Store employee generating nearly $500,000 in revenue. Hundreds of millions of customers visit an Apple Store each year, which is astounding when you consider that there are fewer than five hundred of them worldwide. Go visit one right now. It will be mobbed, with a line to the cash registers more than a dozen bodies deep.

"During a decade of explosion of electronic commerce, a technology store became the world's favorite store," said Ron Johnson, the former VP of retail at Apple, who created and ran the company's stores from 2000 until 2007. At the time, Apple had been selling its products online for several years, but the company lacked a direct connection to the fiendishly loyal fans of the brand and its computers, especially back then, when Apple computers remained relatively niche machines compared to the PCs that dominated the market. Johnson told me that Jobs saw computers sold at such retailers as Best Buy and Circuit City as mere commodities, where they were interchangeable and largely worthless in the eyes of consumers.

"Steve believed Apple was going to win on innovation. The innovation you need to convey to a customer can't be conveyed through marketing," Johnson said. "He knew that if Apple didn't control the point of sale, we'd never get credit for the innovation around the products. There was never a debate whether it would be brick-and-mortar retail." Johnson created stores that were the embodiment of the new Apple brand: sleek, brightly lit boxes of brushed aluminum and polished black glass that felt like the future, but a future with a strong human presence. "Customers need help," Johnson said, especially with computers. "They could go online, they could go to our website, they could find a friend, but they prefer the convenience of going to a store." So, he created the Genius Bar, a next-generation help desk where you could not only get a broken computer repaired, but sign up for inexpensive one-on-one courses in Apple software and hardware. This built an even deeper relationship with customers. Johnson has carried this hands-on touch to his latest venture, the e-commerce company Enjoy, which offers free in-person support for purchases right in your home. A Genius Bar to go, basically.

What's even more amazing about this is that the Apple Store sells the exact same Apple products as every other retailer, but for more money. "It's the most expensive place to buy Apple products," Johnson said, with pride. "They can be bought at a lower price every day, everywhere else," and yet it's the Apple store where the hardcore fans will line up for days before a product launch, sleeping in lawn chairs and living off Doritos, simply to be the first to hold the latest iPhone in their hand, and happily pay more for that privilege. If you looked at it from the cold logic of the market, consumers paying a premium to shop at the Apple Store made no rational sense, but Johnson reminded me that shopping isn't an entirely rational act. "We're people first," he said.

⁂

*T*he new bookstores in New York and elsewhere view themselves in the same light, which is why they all refuse to match Amazon's unbeatable prices. "You can't compete with Amazon," said Christine Onorati, who opened WORD in Greenpoint, Brooklyn, in 2007, and a second location in New Jersey in 2013. "If we look at it that way, we'll lose every time. We can't compete on price or delivery. We have to be a totally different option." Instead, WORD competes with its selection (quality rather than quantity), personal service, events (especially school book fairs), and most important, a sense of desire, not just for the books WORD sells, but for shopping in the presence of likeminded people. "I don't want anyone to think of me as a charity," Onorati said. "No one should think they have to shop at my store, I want them to *want* to shop at my store."

Book Culture's Chris Doeblin was unambiguous about what he wanted to achieve when he opened his third and flagship location on the Upper West Side in December 2014. Doeblin is in his late forties, stands over six feet tall, and dresses in the worn corduroys and thick knit sweaters of a New England fisherman. When we first met he had a pencil behind each ear, and was directing the newly hired staff shelving books. Although he speaks with a soft voice, Doeblin's words are filled with a purpose and also a righteous anger.

"Today, nobody sees books," Doeblin said. "All the stores around here are closed. You take a major reading city like New York and take books away from the landscape—POOF!—they go away." This mattered, because New York was the world's literary capital, and many of the publishing industry's editors and executives lived in the neighborhood. These people needed a place that proved the continued worth of their profession, which had been dragged through the mud in the public consciousness. They needed to see books lovingly displayed in windows, and a place where book lovers could gather and buy books in each other's presence. They needed a place for children to discover books, and for parents to read to them and buy them books. A city without a bookshop to take your kids to was a horror to Doeblin. "We're doing something seriously good," he said. "We know this. That books do ultimate good. That's an idea I can sell."

Doeblin had worked in a number of independent bookstores around New York in the 1980s, and opened Book Culture up near Columbia University in 1997, focusing on textbooks. In 2009, he opened a smaller second location two blocks away (with a more commercial focus), and in 2010, brought in the store's manager, Annie Hedrick, as Book Culture's co-owner. Hedrick is more than ten years Doeblin's junior, talks incredibly fast, and every time I visited the store, her one-year-old son was strapped to her body. Although the second Book Culture location cost $300,000 to open, it quickly achieved annual double-digit growth, and soon surpassed its initial investment. "Looking at that sales and growth," Hedrick said, "we saw that this was possible, and not crazy."

Doeblin began searching further downtown for a third location, and saw the space on Columbus Avenue at 81st Street, when Hedrick's son was just three weeks old. It had a huge storefront window and was a block from Central Park, the American Museum of Natural History, and two subway lines. The rent cost $35,000 a month, but Doeblin felt it was an unparalleled location for both locals and tourists. Better still, the landlords had once operated a bookstore in the space, called Endicott Books, which closed in the 1990s, and was the inspiration for Meg Ryan's plucky independent in the movie *You've Got Mail*, which is forced to close when Tom Hanks opens a big-box bookstore a few blocks away.

Over the following year, Doeblin and Hedrick transformed the space from a furniture shop back into a bookstore. It featured brick arches, different sections nestled into little nooks, and a huge basement dedicated to kids, with separate areas for stories, toys, young adult books, a dedicated cashier, and even a diaper-changing room and small kitchen so parents and nannies could comfortably stay for hours.

In addition to more than $250,000 worth of books, half of the shelves were stocked with other items, from greeting cards and paper notebooks, to gifts, kitchenware, and small garments such as hats and gloves. To pay for all this, Doeblin had emptied his retirement account and mortgaged his family's apartment, basically putting everything he owned into the store. He sent handwritten letters to over a hundred potential investors but only one responded. Rick MacArthur was the publisher of *Harper's* magazine, a staunch advocate for independent publishing, and he made a significant investment in the store. Book Culture's Columbus Avenue location opened a week after Thanksgiving 2014.

To make Book Culture's new store work, Hedrick and Doeblin had to sell over $3 million worth of books and merchandise a year, just to meet costs. This seemed insanely ambitious in an industry that was still defined by decline and increasingly driven by digital publishing. But there were hopeful signs emerging that consumers were not only returning to brick-and-mortar bookstores, but to printed books as well. In 2014, Nielsen BookScan, the largest tracker of sales data for the publishing industry, saw paper book sales grow 2.4 percent from the year before, the first upswing in growth since eBook sales took off in 2010, which was followed by 2.8 percent growth in 2015. At the same time, sales of eBook devices, such as Kindle, Nook, and Kobo, appeared to have plateaued. A 2015 Pew Research Center survey noted a 32 percent decline in the number of respondents who owned an e-reader from the same period in 2014, and the UK bookstore chain Waterstones noted that book sales were up 11 percent over the Christmas season of 2014 compared to the year before, while sales of Nook e-readers were markedly down. A few months later, struggling Barnes and Noble finally turned back to profit as a company, with *Publishers Weekly* crediting the shift to a dramatic decline in money-losing eBook, Nook, and online sales, offset by a relative growth in physical book sales, as well as resurgent brick-and-mortar sales. That doesn't mean that digital books won't continue to grow. They may well. But the heady predictions that eBooks would do to paper books what the MP3 did to physical music look increasingly unlikely.

Part of this can be attributed to readers returning to print. I was once a bookstore fiend, buying dozens of books a year from every type of bookstore. Then I started ordering from Amazon and I found myself less involved with books and reading than I once did. It's not that my love of reading vanished. But a part of reading's charm was lost when the acquisition of books moved online, just as it had for music. Then I bought my first Kindle. Initially, my love returned. Here was the world of reading in the palm of my hand, with anything I wanted two taps away. It weighed nothing, stored infinite libraries of information, connected anywhere, and lit up, so I could read in bed at night. I devoured titles on the Kindle, while my consumption of physical books ground to a halt.

But after a few years, I fell back to print. I'm not sure what drew me to it, but a couple of factors contributed, including many that I discussed in the last chapter. I joined the public library, and because of this I started reading paper books (largely for work). I quickly found how much I actually missed holding a book in my hand and reading from paper. It was a vastly superior experience, for reasons that seemed counterintuitive, when the technology of my Kindle had so many obvious advantages. Yes, a book was heavy, but I knew where I was by the very feeling of the book's thickness between my fingers, something that I desperately craved on the Kindle. I couldn't annotate to the cloud as I read in print, but I could underline, write notes, fold down corners, and never get lost by accidentally tapping the page with my finger. With paper, I couldn't enlarge text or turn up the backlight, but I could read without having to charge a battery. I could accidentally step on a book and not have to pay Amazon $140 for a replacement. Today, I get books at the library and bookstores, borrow them from family and friends, and have a pile on my nightstand. I only use my Kindle when I travel for longer than a week. The rest of the time it sits in a drawer, its empty battery logo pleading for a charge.

There are numerous economic factors behind the revival of bookstores—the opportunity Borders created when it closed, the postrecessionary recovery—but something deeper seems to be at play. Elizabeth Bogner, one of the managers at Book Culture, noted this to me as we watched a stream of people steadily enter the store, find the nearest staff member, and express thanks for opening. "I'm seeing a community who, having lost bookstores, now understand their full value," Bogner said. "The community went through its five stages of grief. They said, 'Who would be stupid enough to open a bookstore?' But then, it happened."

"There's an in-place premium" to shopping at Book Culture and other bookstores, Doeblin said, as we watched the new nonfiction table swarmed by shoppers scanning, reading, and flipping through new titles. "We can sell a book off the shelf with no regard that it sells for a penny on Amazon. People will pay for books they see." That is because books are an aspirational consumer product, especially today, when so much reading and time are spent online. "If you spend twenty-six dollars on a book, you're aspiring to stimulate your intellect, to get involved with literature. You've got a lifestyle that affords intellectual curiosity. Only the highest-level consumer is buying and reading books today," Doeblin said. "We're talking about the richest, best-educated consumers, the most coveted in retail, and they should be cherished like gold. Books are the apex of the consumption pyramid!"

Shopping transcends our need for consumption. The pursuit of goods is an excuse for social interaction. The conversation that happens in the store is vastly more important than whatever *shmata* or tchotchke we buy there. This is true at the weekly market in a remote village in New Guinea or on a Saturday afternoon at the massive Apple Store on Fifth Avenue, where tourists from every nation on earth are fondling iPhones. We are hardwired to shop. It is how we entertain ourselves. When we get together with friends on the weekend, we go to the mall, window-shop, browse, peruse, and check out stores. Bob Grayson told me that one and a half days of an average American's weeklong vacation are spent shopping. None of that social interaction can happen online, regardless of how well you design social media plug-ins, or how many unboxing videos you view on YouTube.

I don't consider myself a shopper. My wife describes my presence in a clothing store as a wet wool blanket draped over her shoulders. But place me by the window of a record shop, a bookstore, or God forbid a modern furniture boutique, and I start spending like a Rockefeller. This is especially true at markets, whether bazaars in the Middle East, food halls in Europe, or urban American markets such as the Brooklyn Flea, which remains one of my favorite shopping experiences.

The Brooklyn Flea has grown from a small flea market set up in a schoolyard in 2008 to a model of entrepreneurial analog capitalism emulated the world over. Its premise is basic: an urban, young market with a mix of older merchandise (vintage clothing, furniture, trinkets) and new (artisan food, screenprinted T-shirts, art). It is a gathering place for the community both physically and emotionally associated with it.

"Our markets aren't commerce, they're hangs," said Eric Demby, the Brooklyn Flea's cofounder. "It's outside, it's free, you don't have to buy anything, and you will pretty much run into someone you know." I have bought some cool things at the Brooklyn Flea, but I always ended up there because it is a fun place to spend a few hours on a weekend. Demby said that while many of the Brooklyn Flea's vendors also sell on Etsy, the standardized format of the online craft marketplace makes it practically impossible to stand out from the five hundred other people selling typewriter-key jewelry. "The Internet will never have nice locations," Demby said.

The Internet and its retailers face a far greater challenge earning a customer's trust than brick-and-mortar retailers. Last winter, I ordered an obscure book on real estate investing from a third-party seller on Amazon, because it wasn't available anywhere else. A few weeks later, the seller contacted me, told me he'd be in Toronto, and inquired whether he could drop the book off at my house, because it would be "easier" and "cheaper." Why not?

A few days later, he wrote back, and told me it would be better if, in fact, we could meet somewhere downtown, because it would be "safer." Safer? What was he talking about? This was a book, not heroin. I told him to drop the book in my mailbox, and when he didn't respond to me, I contacted Amazon customer service, sure that I was being scammed. The shipping cost for the book was now listed as $38, and the seller had no other sales in his record. My paranoia took over, and I imagined a gang of bookselling thugs staking out my house, stealing my identity, and possibly mugging me in some downtown alley, in a book deal gone bad. After an hour attempting to explain my concerns to the Amazon customer service representative in Manila, the company filed some sort of complaint with the vendor. The next day, the book appeared in my mailbox, and I received a nasty e-mail from the vendor, calling me rude, ungrateful, and stupid, because he went out of his way on a business trip to deliver this book, and I should now pay for his extra taxi costs. Because all our correspondence was filtered through Amazon's system, and the vendor and I remained anonymous, our thin veneer of trust eroded when our communication broke down. If we could have spoken directly, or conversed face-to-face in a store, the whole thing could have been sorted out from the start. Instead, miscommunication led to misunderstanding and we both felt ripped off.

An e-commerce merchant cannot engage deeply with you as a person, let you taste a sample, share an honest opinion, or flirt with you. E-commerce is a platform for delivering goods and services, no more. While online such retail platforms as eBay, Etsy, Craigslist, Amazon, and others refer to themselves as communities, with conversations and interactions between buyers, vendors, and creators happening around the world, the strength of those retail communities pales when compared to that of a brick-and-mortar retailer. When Lexi Beach opened the Astoria Bookshop in Queens, New York, in 2013, local residents provided her with funding, painted walls, installed shelving, and unpacked boxes without asking for a penny. "Strangers showed up in their own spare time to make the bookstore open faster than it would have otherwise," she said. "It felt to me like this was the universe rolling out the red carpet in front of me, saying, 'This is the thing that needed to be done.'"

At Book Culture, when a knitting club asked Doeblin whether it could meet there regularly, he not only said yes, but offered to supply the knitters with tea and cookies. When a young customer told his mother he wanted to have his bar mitzvah reception at the store,

Doeblin wished him mazel tov and helped the family plan the party at no cost. Sure, these acts of goodwill might pay back in sales from these customers, but most likely they wouldn't. They had a deeper purpose.

"When small stores are replaced by chains and Amazon, what do we lose?" asked Antonin Baudry, France's cultural counselor in New York and the author of several graphic novels. "We lose something specific. It's called a city." A few months before Book Culture's own opening, Baudry had opened Albertine, the most lavish, beautiful gem of a bookstore in New York, located in the Cultural Services division of the French Embassy, across from Central Park. A city, in Baudry's definition, was a collection of businesses, such as bookstores, which paid taxes, allowed citizens to meet, and ultimately contributed to the cultural and physical landscape around it. "If all that goes away, it's not a city anymore."

———————

*O*n my last night in New York, I attended Book Culture's store-opening party. For five hours you could barely move amid the shelves, as nearby residents, publishing industry figures, friends and family, an old golden retriever, and eager book lovers packed the place. There was wine and cheese and cake. Best-selling young adult author Tim Federle mixed cocktails and signed copies of his alcoholic nursery rhyme book *Hickory Daiquiri Dock*. At one point, Doeblin stood behind a lectern at the rear of the store and addressed the crowd. He spoke about the worth of books, culturally and as assets, and the need for the publishing industry to wake up and support those advocating on their behalf, including such stores as Book Culture.

"We need profitability in a space like this," Doeblin said, beseeching the crowd to return often and vote with their wallets to create the store, the neighborhood, and the city they wanted. "We'll be here as long as you guys come."

Over the next months, they answered his call. Book Culture had a great opening season, growing its sales even more than expected throughout the spring of 2015. Its contemporaries continued to grow as well. Greenlight and McNally Jackson have both expanded to second locations they opened in the past year, while WORD, Astoria Books, Albertine, and other resurgent independent bookstores nationwide continued steadily to sell more books, without an end in sight. When I last spoke with Doeblin, he was already busy looking for a location where he could open his fourth store.

7

The Revenge of Work

West Canfield Street, located in Detroit's Midtown neighborhood, is a standing refutation of the much rumored death of the old ways. The gentrified strip boasts a craft beer pub, an espresso bar, a cold-pressed juice counter, a clothing boutique, and two adorable gift and home stores that carry "Smile, You're in Detroit" mugs, T-shirts, and tea towels. At the center of this is the grandest store of them all, the flagship retail location for Shinola. Here, the exposed steel rafters, artfully faded brick walls, and polished concrete floor lend an airy atmosphere of industrial hipster elegance, meant to evoke the look and feel of Shinola's products.

When I visited, these included hand-stitched leather baseballs, basketballs, and footballs ($40–$225), a wooden screwdriver set ($65), linen-covered journals ($12–$20), and beach towels ($160). There was a wall of pet accessories, ranging from dog-shaped cushions to leather leashes and collars, and another display of leather wallets, smartphone cases, and handbags. Behind the cash registers, bearded men assembled bicycles with steel frames handmade by members of the Schwinn family ($1,000–$2,995). The store had cool music, friendly, attentive staff, and a coffee bar where you could sip a cold brew on distressed leather couches under a framed American flag stitched by military veterans ($15,000) and read a copy of the independent dog magazine *Chewed*.

The main attraction was Shinola watches ($450–$1,200), assembled in a factory just a mile away. They were displayed around the store, laid out on leather-wrapped trays and in glass cases. Shinola watches have a classic, utilitarian look, like something out of a railway station or the golden age of Detroit's auto business: heavy, elegant, masculine hunks of chrome and glass. They come in six different models and dozens of styles, from the minimal Arabic numerals on the women's Birdy to the beefy Black Blizzard chronograph. Shinola watches are all completely analog. They don't connect to a phone or count your steps. They just tell time.

I was visiting Detroit a year and a half after Shinola had sold its first watch, and less than two since they began to assemble them nearby. When the company's founder and owner, Tom Kartsotis, opened this retail store, he figured it would sell maybe $180,000 worth of watches over six months. Instead, Shinola's Detroit store sold more than $3 million in goods during the second half of 2013, and more than $9 million in 2014, mostly in watches. The company now has other boutiques open in New York, Minneapolis, Chicago, DC, Los Angeles, and London, with more coming soon, and its products are sold at more than a hundred other retailers, from small men's clothing stores to Nordstrom and Saks Fifth Avenue, and online. Shinola watches are everywhere. Bill Clinton apparently owns a dozen.

Although Shinola products have the heritage look of a decades-old brand, the company is brand new, and no scrappy startup. Kartsotis is the founder and former chairman of Fossil, the watch and apparel company that dominated shopping malls in the 1990s, which is worth roughly $3.5 billion today. Kartsotis's personal wealth is conservatively estimated at hundreds of millions of dollars, and Shinola has been his largest project since he formally left Fossil in 2010.

Shinola initially began as Kartsotis's plan to build a domestic private-label watch factory that could make watches for such customers as Tiffany and Movado. When Kartsotis, who resides in Texas, was looking for a location, Detroit was an obvious option for the factory, thanks to the city's affordable factory space and a workforce with manual skills. But Kartsotis wasn't sure whether people would purchase a luxury brand made in Detroit, so he commissioned a survey that asked consumers whether they preferred to buy $5 pens made in China, $10 pens made in America, or $15 pens made in

Detroit. Respondents overwhelmingly chose the most expensive pen, simply because it was made in Detroit.

Kartsotis realized that Detroit had untapped luxury potential, and a "Made in Detroit" watch brand had much more upside than a factory making watches for other companies. When Kartsotis told a friend in the watch industry he planned to make high-end watches in Detroit, the friend reportedly said, "Tom, you don't know shit from Shinola," a popular phrase among soldiers in World War II, who were referencing a brand of shoe polish. The name stuck.

Shinola's entire brand rests on its location in Detroit. Its motto is "Where American is made" and the company touts the domestic provenance of all its goods, whether they have been created and assembled at the Detroit factory or by suppliers in other states. The Shinola marketing material is relentless in pushing this narrative of American artisan craftsmanship and ingenuity. Photographs and stories of proud Shinola workers decorate its stores and anchor the company's advertisements. The Shinola website features a glossy video of workers assembling watches, as sunlight pours in through the windows and the rugged voice of a worker intones: "This is the city that made this country. With its steel, with its skill, with its labor. It's why we're here. Reestablishing trades that haven't been seen in this country in a generation. Knocking rust off the American supply chain wherever we can."

When the camera pans out to reveal a parking lot filled with Shinola workers, you're almost ready to chant "USA! USA! USA!"

This scrappy can-do message, married with Kartsotis's disproportionate wealth, has made Shinola a magnet for critics. The *New York Times* published a scathing review of the company's TriBeCa store, a multimillion-dollar boutique in a building the Kartsotis-owned Bedrock Manufacturing purchased for $14.5 million. The author called Kartsotis a "midprice watch mogul looking to go luxury under the cover of charitable business practices" whose attempt to sell expensive watches with a handcrafted, bootstrap-origin story was a bait-and-switch. "Shinola is a self-described luxury brand animated by do-gooder impulses, taking the conceit of 'Made in Detroit' and squeezing every last retail penny from it." Pure carpetbaggery.

Locals had their own complaints. Shinola was profiting off the city's hardship, tugging at the heartstrings of Americans to help support poor Detroit, whose ruined, abandoned buildings had become a form of postindustrial pornography to those living elsewhere. Perhaps an even greater insult to Detroit was the notion that these made-in-Detroit goods were entirely unobtainable to most residents of the city Shinola was waving its $15,000 flag for.

Shinola has done little to avoid this. Shortly after the City of Detroit declared bankruptcy, succumbing to crushing debts that had decimated city services to the point where its residents lived without water, fire engines were held together with duct tape, the famed Detroit Institute of Arts contemplated selling off its masterpieces, and numerous Detroiters actually froze to death in their homes, Shinola's Twitter feed displayed the following message: "Bankruptcy, shmankruptcy. We have a lot of job openings here in #Detroit. Come join our team!"

It may be fair to criticize Shinola on these grounds. But take note of what's being implied here: a capitalist mogul is accused of acting greedy. The very idea that there was a business opportunity in Detroit, no matter who seized it, was supposed to be unthinkable. Shinola is not run by a culture warrior in pursuit of a personal goal. Tom Kartsotis is a businessman who saw going analog as a way to make money.

Shinola may base its brand on a fanciful tale of renewed American manufacturing, but the dollars generated by the sale of its watches, and the jobs those sales have created, are undeniably real. The benefits of these jobs, and the business model of Shinola and other analog industries, have tangible, long-term benefits for investors, workers, and communities, which differ greatly from those created in a digital economy, whose own benefit is far less widely spread.

———————

The digital economy is a broad, wildly imperfect term. It came about during the last tech boom, in 1995, as the title of a best-selling book that pointed out how the growing use of the Internet would fundamentally change business. Digital work spreads across pretty much every industry, from computer software and hardware companies to divisions within traditionally analog firms that focus on digital tasks, such as e-commerce and information databases. Other terms have been used as synonyms for the digital economy, including the knowledge economy, information economy, Internet economy, and the utopian-sounding new economy.

The core idea is that digital technology is a transformative force that can deliver vastly more efficient products and services to consumers at a lower cost, and with greater ease, across time and space, in ways that traditional analog industries cannot compete with. The digital economy is disruptive. It upends markets and dispels long-held assumptions about business. In the industries where digital technology dominates, this has proven true, and the impact of digital technology on the greater global economy has been tremendous. It ranks up there with the invention of steam power, electricity, and telecommunications. But it also contains an inherent assumption: that analog economic activity, and its associated work, will gradually be replaced or simply disappear.

The commonly stated goal behind the digital economy is the maxim of creative destruction, coined by economist Joseph Schumpeter in the 1950s to describe revolutionary industrial change that consumes old processes and ways of doing business. As the hundreds of thousands of people who once worked for Kodak can tell you, this destruction is real. Wherever the digital economy has staked a claim, analog incumbents have struggled to adapt. The narrative of the digital economy's creative destruction truly blossomed during the expansion of two forces in the 1990s: the first major commercialization of the Internet through web browsers, and the rapid globalization of a post–cold war world, where American neoliberalism became the dominant economic and political philosophy. It was a marriage made in heaven. The capitalist optimism of American economic and technological progress was irresistible to entrepreneurs the world over. Citizenship now mattered less than education and connectivity. Anyone with a computer and modem could now compete on a global scale. Your garage was just as powerful a corporate headquarters as an office tower in Manhattan or Tokyo.

This was the storyline pushed by the stakeholders of the digital economy, which included business leaders, free-market economists, politicians, and the media. A cheerleader that stands out for me from this time is the mustachioed *New York Times* columnist Thomas Friedman, who extolled the virtues of the new global economy in his articles as well as his best-selling books *The Lexus and the Olive Tree* and *The World Is Flat*. Friedman always seemed to be talking about meeting a new friend, the young, Western-educated prince of some formerly poor nation, who was in the process of transforming that country by rapidly expanding the Internet, or shifting the economy from farming to writing software for American companies, in a process known as outsourcing. If Americans didn't rethink their old ways and plug into this economy, Friedman always concluded, they would be left in the dust. The solution was simple: more education and entrepreneurship. If Shen in Shanghai could start a billion-dollar T-shirt company out of his apartment, then what would stop Suzy in Scranton from doing the same, using the same factories in Vietnam, while handling customer service virtually through a call center in Bangalore?

There was a powerfully postindustrial romance running through this. With the aid of digital technology, we were now free to shift the focus of our economy from outdated, broken, dirty industries and professions (manufacturing, resource extraction, manual labor), and focus on the good jobs of the future, which were defined by information and creativity. In the West, we could devise the best products and services, and have them made, with the aid of teleconferencing and broadband connections, by those in other countries who would happily do our dirty work for less, raising both our standards of living. Put down that shovel! Throw out your wrench! Pick up a mouse and make a website!

If Netscape, Microsoft, and outsourcing defined the first round of the digital economy, then the current version is defined by Facebook, Apple, and automation. It is an advanced and accelerated version of the same creative destruction from twenty years ago, but on a larger scale, with much more computing power in the hands of billions. The digital economy has become even more attractive, especially in the wake of the Great Recession, which decimated traditionally dominant businesses such as manufacturing, real estate, and finance. While old industry giants such as General Motors and General Electric were pandering for bailouts, companies such as Twitter, which counted their staff in the dozens, were being valued at many billions of dollars. Why invest in a blue-chip company struggling to adapt, when a small investment in a tech startup could make you rich overnight?

Today, it is the titans of technology—Tesla's Elon Musk, Facebook's Mark Zuckerberg, Uber's Travis Kalanick—who are the new gods of capitalism. Their stories of rapid success are the subjects of best-selling biographies and Hollywood movies. Silicon Valley has supplanted Wall Street as the destination for the best and brightest. *The Economist* reported that in 2014, one fifth of American business school graduates went to work in technology.

None of this has been lost on politicians. They pepper their speeches with talk about innovation, and sign bills to fund community Wi-Fi zones, grants to digital startups, research hubs, and technology incubators. The reason behind this is jobs, they say. Not just any old jobs, but the jobs of the future. Jobs that can compete globally, and create more jobs going forward. Jobs that are creative and solve real problems. The good jobs.

The reality is a lot less simple. While the growth of the digital economy is real and will only continue, the benefits of that vast growth on employment, economies, and communities have not even come close to matching the hype surrounding them. Those other jobs, the ones politicians and thought leaders don't talk about—analog jobs—still matter a hell of a lot more than do those associated with the digital economy. Nowhere is this clearer than Detroit.

---

*D*etroit *is compelling for those* studying the American economy because its wounds are out in the open. Silicon Valley has some creeping problems—most notably very high levels of homelessness due in part to its pricy housing stock—but they are little known. Detroit's struggles, on the other hand, are visible the second you drive into the city: vacant lots, crumbling factories, and ghost buildings. To many, Detroit is a graveyard for the predigital American capitalism of the twentieth century. Technology is not at the heart of Detroit's decline, which has its roots in racism, the auto industry's and organized labor's combined mismanagement in the face of global competition, and staggering political incompetence and corruption. But the digitally enabled globalization that impacted so many American businesses in the 1990s took hold in Detroit years before.

The car companies pioneered the use of automated robots to replace human labor, and embraced outsourcing and offshoring long before India-based call centers became a popular phenomenon. When the Great Recession hit, it hit Detroit and the surrounding region harder than it did the rest of America. The real estate bubble's collapse erased the modest wealth of many Detroiters, whose homes were their only asset, and the decline in consumption and stock prices squeezed auto manufacturers past the breaking point. Car plants closed and slashed shifts. Parts suppliers did the same. Workers from those factories lost jobs, wages, and benefits, and fell further into debt. Detroit's meager tax revenues pretty much disappeared. The city's population shrank even further. Unemployment in Detroit soared to a peak of nearly 30 percent.

The Great Recession was most directly caused by too much bad debt, but many believe its roots lie with digital technology. While global asset bubbles are nothing new—the Dutch tulip crisis occurred centuries before computers—the rash of reckless lending that led up to the asset crash of 2008 was directly correlated with the integration of high-powered computer algorithms and the financial markets. Software allowed bankers to divorce actual assets (homes and property) from the circumstances and risks associated with them. Loans were issued to people for houses, but the banks buying and packaging them together into securities had no idea where those homes were located, who owned them, and what the underlying risk of the loan was. Software based on past data set prices for these loans and assets using inaccurate economic assumptions (for example: housing always goes up, because it has the past 20 years). Many in Wall Street bragged how computers had basically eliminated risk from investing.

Douglas Rushkoff, an author on technology, said this created a form of "hypercapitalism" utterly divorced from the laws of supply and demand, actual commerce, and the creation of value. The speed and volume of transactions that digital technology enabled amplified the volatility of the market and the asset bubble more than anyone could have predicted. "The limitations of organic human memory and calculation used to put a cap on the intricacies of self-delusion," wrote computer scientist and philosopher Jaron Lanier in his book *You Are Not a Gadget*. The rise of quantitative hedge funds turned capitalism into "search engines," which Lanier described as the perfect "blending of cyber-cloud faith and Milton Friedman economics." The recession hit when the real world couldn't support the numbers being assigned to it by computers and speculators. The computer's calculations were off, it turned out. House prices can go down. Pop went the world, and with it, Detroit.

Unemployment in Detroit has since fallen from the nauseating heights of the recession, but it is still among the highest in America, and Detroit remains one of the nation's poorest cities. Detroit needs jobs, which is the promise that Shinola has staked its brand on.

———•———

S hinola's headquarters and factory occupy an entire floor of the Argonaut building in Detroit's Midtown neighborhood, which originally served as the research and design division of General Motors. Some five hundred people work for Shinola, and many of those jobs involve the assembly of watches in its Detroit factory. While Shinola began assembling watches from components bought from other suppliers (leather straps from Florida, watch components from Switzerland, Thailand, and Taiwan), increasingly the company has been moving production of its parts in-house. In 2014 Shinola began manufacturing its watch faces and leather straps in Detroit.

Paloma Vega took me on a tour of the leather strap production line, which she ran. Initially Vega (who previously worked for Louis Vuitton) hired people who had experience in automotive textiles, because the skills gained from stitching leather seats were not dissimilar to what it took to transform a tanned cowhide into a watch strap. "It looks very simple," Vega said, as she walked me through the process, "but it is not." Although the tasks seemed repetitive, it required constant adjustment to create a consistent strap. Each piece of leather was completely different in its look, texture, and density. "With leather you always need the human touch," she said. "You cannot set up the same line with only machines and get good results."

There were some forty people in the leather division, and they worked at various machines in a spacious loft space flooded with soft, natural light, and a soundtrack of soul and Motown hits. Workers regularly rotated to different machines, so as to learn the entire process. Most assembly employees were African Americans in their twenties and thirties who had worked in the auto industry, and everyone wore some item of Detroit clothing and a Shinola watch that the company gave them. Tarez Franklin was the newest employee, just days into his job of punching holes into the straps. He had a communications degree, but had been stamping out mufflers at a Chrysler plant before being laid off. "I like the hands-on aspect and the movement here," Franklin said. Muffler assembly was entirely automated, but this job required more decision making. "Here, with every part, there's more of the human touch. It makes you feel like you're really part of the company."

Manufacturing has acquired a reputation as a dead-end job, free of independent thought or creativity. That goes back to Henry Ford's assembly line, and the innovations in workforce management that followed it, which sought to strip every possible human variable from manufacturing to increase efficiency. But Shinola was pursuing an approach called "skill at scale," which was similar to the "industrial/artisanal" model Nicola Baldini hoped to build at FILM Ferrania. This sought to marry industrial efficiency (the leather line made around 400 straps a day) with the handmade touch that gave a product its added value. The challenge, according to Vega, was to undo the classic assembly-line mentality most of the workers had grown up with, and make them understand that they were an active and dynamic part of the manufacturing process.

The term economists have used for this is *reskilling*, which is the antidote to the phenomenon of deskilling that has been a natural consequence of automating workflow. In his excellent book on the cost of automation, *The Glass Cage*, Nicholas Carr defines deskilling: "As more skills are built into the machine, it assumes more control over the work, and the worker's opportunity to engage in and develop deeper talents, such as those involved in interpretation and judg-ment, dwindles. When automation reaches its highest level, when it takes command of the job, the worker, skill wise, has nowhere to go but down." Carr tracks the consequences of this, from plane crashes that occurred because pilots became too reliant on autopilot to doc-tors who miss diseases because they use diagnostic software. The most common example of deskilling I see involves Uber drivers who blindly follow their GPS guidance, even while I am shouting from the backseat that I can see the address out the window.

Reskilling seeks to return human judgment to the automated workplace. "It's our ability to make sense of things," Carr writes, "to weave the knowledge we draw from observation and experience, from living, into a rich and fluid understanding of the world that we can then apply to any task or challenge. It's this supple quality of mind, spanning conscious and unconscious cognition, reason and inspiration, that allows human beings to think conceptually, criti-cally, metaphorically, speculatively, wittily—to take leaps of logic and imagination." While this sounds high-minded, it doesn't only apply to artisan industries, such as watchmaking or food. Over the past few years Toyota, once a pioneer of automated manufacturing, has been replacing some robots with human workers at assembly plants across Japan, to develop new skills, improve manufacturing processes, and ultimately build better cars.

On the other side of the Shinola factory was the watch-face as-sembly division, which operated in an opposite universe from the leather line. The lighting was bright and fluorescent, and there was no music or even talking, because open mouths projected spit and other particulates, and you couldn't afford having minute pieces of some-one's lunch land amid the gears of a watch. Everyone wore blue lab coats and protective netting over their hair and shoes to keep dust and contaminants away from the precise mechanisms. The workers were bent low over their stations as they snapped miniscule dials, gears, and components together with tweezers, or twisted screws smaller than a grain of rice into place.

Supervising the process was Willie Holley, a former security guard in the building with no previous experience in manufacturing, who subsequently became the face of Shinola when his portrait accompanied the company's first ad campaign. Holley had begun on the same line just two years back, assembling these tiny parts, but had been promoted several times, with corresponding increases in pay and benefits. "This isn't the end of the line for these people," said Jennifer Guarino, the VP of leather at the company. "If they excel we need masters to teach and to supervise. We get criticized and people say 'Well, you only give them $12 an hour,' well, that's a good wage, and a step on the ladder." The factory's structure was based on the successful German Mittelstein apprentice system, which taught skilled trades to factory workers and allowed them to grow within the business. Kartsotis told me that $12 may be the starting hourly wage for trainees at Shinola, but jobs here include full benefits after several months, and regular salary increases, which make the average wage at Shinola's factory somewhat higher.

Many critics of Shinola argue that the company romanticizes jobs that shouldn't be romanticized. If the economy's future is in digital technology, then why wave the hand-stitched American flag and herald the return of domestic manufacturing, when it was an overreliance on manufacturing that got Detroit into this mess? Wouldn't that investment be better spent training these people to write code or design websites? The problem, according to Guarino, was one of perception. All the talk about knowledge work and the digital economy had saddled manual work with an unsavory reputation. "We don't have a skills gap in America," she said. "We have a value gap. If you don't value something, people don't want to do that. There's inference that you haven't done enough with your life if you're a pipe fitter. That's the wrong valuation."

Amy Haimerl, a journalist with *Crain's Detroit Business* who had moved here from Brooklyn in 2012, praised Shinola. The company was creating good, well-paying jobs in a community where the skills, experience, and DNA of the workforce were geared toward "making shit." To those few Detroiters who managed to secure one of those jobs at Shinola, it was the manufacturing equivalent of getting hired by Google. "We still need physical objects and people to produce them," Haimerl said over a beer at PJ's Lager House, a dive bar with a dive record store in its basement. Haimerl was the only one in her family to go to college and she earned a decent living, especially by Detroit standards. Her brother, on the other hand, was a diesel mechanic who could fix anything, and had to shoot coyotes and scavenge roadkill just to feed his family. "That's fucked-up," she said, downing a shot of Jameson. "We put our value in technology jobs and forget those who built things. We treat them as inferior. If we continue that, our country is fucked."

———◆———

The facts are on Haimerl's side. In America and other industrialized nations, job creation has been steadily declining, along with real wages (on an inflation-adjusted basis). The GDP per person is down, inequality is up, and labor's share of capital (the percentage of GDP that goes to workers) has steadily eroded. For all its wealth creation and gains, the digital economy, as it stands, has not delivered any substantive gains in employment and wages. Technology is surely not to blame for all of this, but it does play a significant role.

Digital technology, as an industry, is a relatively paltry creator of jobs. Tens of thousands of people may work at such companies as Google, Facebook, and Amazon, but those numbers are a drop in the ocean when you compare them to what an analog business, for instance the Ford Motor Company, employs. The lean nature of technology companies is inherent in their model. They can start in a dorm room with one person and scale rapidly, reaching millions or even billions of customers, without having to grow the equivalent infrastructure of physical factories, warehouses, and stores and to staff those with human beings requiring salaries and benefits.

Computers allow work to be done with relatively few people, if any at all. A data center can take up dozens of acres and cost a billion dollars to build, but requires just a handful of employees. This is how a firm such as Instagram can become the leading photography company in the world in just a few years, worth a billion dollars at the time of its acquisition by Facebook, with roughly the same number of employees as a store like Book Culture. The only US technology company to even crack the country's top twenty employers is Hewlett Packard (HP), and its workforce has shrunk drastically in the past few years. HP has laid off more than fifty thousand employees since 2013, the equivalent of everyone who works for Google.

"There is no economic law that says that everyone, or even most people, automatically benefit from technological progress," wrote economists Erik Brynjolfsson and Andrew McAfee in their ground-breaking 2012 book *Race Against the Machine*, which highlighted the growing gap between technological progress and job creation. "The threat of technological unemployment is real." Brynjolfsson and McAfee are not technophobes gripped by a fear of progress. They point to previous disruptions in labor during the technological leaps of the industrial and mechanized ages, and show how these eventually led to greater middle-class wealth and job creation, as productivity increased. The difference now is the speed and scale of digital disruption, which has accelerated exponentially, and the fact that all of that digital progress has not resulted in any real gains in productivity. The creative destruction so beloved by the technology industry is destroying jobs far more quickly than it can create them.

One of the factors Brynjolfsson and McAfee credit for this is the so-called superstar nature of technology companies. The digital technology industry tends to be dominated by monopolies with few, if any, competitors. The nature of consumption of digital products favors standardization and builds itself around dominant platforms. For computers to talk effectively to one another, they need to speak the same language and use the same formats. At the early startup stage, there may be dozens or hundreds of fledgling companies vying to establish this new standard, but over time, consumers and investors reward only the largest one with tremendous success, while the rest simply fall away. There may be other search engines out there, but Google is the default, the most trusted, the one most used. The cost of challenging a Google is so incredibly astronomical, and its market share is so complete, that it would be nearly impossible to do so from scratch. Why even bother? (If you don't know the answer, maybe you should Bing it.) The same goes for Microsoft's control of PC operating systems and Facebook's imperial command of social networking. Even computer hardware is monopolistic, from Intel's dominance over processors to Foxconn's lock on device assembly. For all its open platforms and democratized credo, digital technology is a remarkably winner-take-all industry. Either your company becomes a billion-dollar "unicorn" or it burns through its cash and disappears in a few years.

This monopoly effect creates significant employment opportunities for those who can get a job at one of these dominant companies, but not exactly a robust market of competitors who create more jobs. Contrast that with an analog industry. While General Motors and Toyota are the world's largest car companies, each only controls slightly more than 10 percent of the market, and the rest is split up among all the other manufacturers, from Volkswagen and Chrysler to Tesla, Ferrari, and Kia, each of which employs people to build its cars, dealers to sell them, mechanics to fix them, and suppliers to make and distribute parts.

The more important reason behind the digital economy's failure to create significant jobs is that minimizing the use of human labor tends to be one of its fundamental goals. "As intelligent machines become cheaper and more capable, they will increasingly replace human labor, especially in relatively structured environments such as factories and especially for the most routine and repetitive tasks," Brynjolfsson, McAfee, and the journalist Michael Spence wrote in an article for *Foreign Affairs*. "To put it another way, offshoring is often only a way station on the road to automation." What began on the factory floor has edged up to the cubicle and corner office, threatening not just warehouse workers and delivery drivers with robots and drones, but lawyers, radiologists, and newspaper reporters, whose jobs are increasingly being done by artificial intelligence software. In a few years, as you step into your self-driving car, what job will it take you to? If the answer is a job that can either be done by a computer with artificial intelligence or a robot equipped with one, you might want to think about a new career path.

Futurists and technology leaders offer a sunnier view: all this displacement could simply be a way station to a period of technological bliss. John Maynard Keynes, the father of modern economics, predicted in 1930 that we would reach the point where machines do all the work we don't want to do, and man simply spends his time basking in idle thought. We will welcome our Terminators with open arms as they fix us cocktails. Or, for those who don't want to spend all day philosophizing, disruption will only create more opportunities for employment. Digital technology has already made many new jobs possible, from systems engineers and application designers to Etsy artisans and Uber drivers. The nature of work has always changed, and it will always change. Perhaps if we are all creative enough, the lowered cost of living (thanks to improved technology) and the rapidly expanding opportunities for on-demand freelance work will more than supplement the disruption of the traditional workplace.

This is a promising tale, but there is a catch. Undoubtedly the technology industry creates many jobs, but the jobs it provides tend to be rather specific and geared toward educated, upwardly mobile (and overwhelmingly male) individuals. Uber drivers notwithstanding, most of these jobs require very specific skills, and the reality is that acquiring those means access to technology (computers) and education, plus a degree of technological literacy that provides a high barrier to entry. Everyone may want to work in digital technology, and the jobs may be there, but it's not as simple as walking up to the doors of Google and saying you're smart, strong, and willing to do whatever it takes. The good-paying jobs in technology are at the top of the food chain, and are realistically accessible only to the small portion of society, the best and the brightest, who already have their pick of plum jobs.

"Not everybody has the skills to be a computer programmer," said economics professor Charles Ballard of Michigan State University, who grew up in Detroit. "I know we lost half a million manufacturing jobs in the past decade in Michigan. What happens when a fifty-seven-year-old man gets laid off from his auto job? You're going to tell me he's going to become a web engineer? You may as well tell me he's going to become an astronaut."

Digital technology has proven very good at creating two types of jobs: high-paying, highly specialized jobs at the top (such as software designers and CEOs), and low-paying, low-skill jobs at the bottom (such as Foxconn phone assemblers and Amazon warehouse fulfillers). The result is an economy of increasing inequality. Across the Western world between 1992 and 2010, the share of employment by middle-skill workers has fallen dramatically, even as high-skill and low-skill jobs have grown. "For a long time, from the end of WWII to the 1970s, we had a lot of high paying jobs that didn't require college education," said David Autor, an economics professor at MIT. "We have many fewer of those. They have been automated and offshored. That means the set of jobs available to people without a college education is less good." You can work at Facebook's headquarters for a six-figure salary as a programmer, or you can clean toilets there for a fraction of that. There isn't much in between.

We tend to view the successes of Facebook and other huge technology companies as the rule, rather than the exception. This feeds into the notion that the solution to the employment problem is simply a matter of entrepreneurship. If everyone goes and creates their own tech startups, that will create the jobs our society needs. "One fallacy in the argument of Thomas Friedman was the passive view that everybody can be brilliant and creative in their own way," said Alan Blinder, a professor of economics at Princeton University, former adviser to President Clinton, and expert on outsourcing. "There are only a few jobs in certain economies for that. Any economy has a lot of routine jobs. And many of those will be low paying." Unfortunately, as much as Suzy in Scranton may want to create the next great billion-dollar app, the reality is that she probably doesn't have the capability or desire to do so. Most people just want to earn a living. Even in the best-case scenario, the market only has the capacity to support a limited number of new technology companies, which by their nature will still create a relatively small number of jobs, involving a small slice of the population.

The growing, digitally powered, on-demand economy may offer some opportunity for piecemeal work, but it is limited in size and scope. As someone who has always worked as a freelancer, I know that it is not a style of employment that is suitable for the majority of work-ers. People need and want stable, well-paying employment, with the hope of improved wages and conditions. As the on-demand economy grows, increased competition from other freelancers will drive down wages for those driving Uber riders or delivering groceries for Insta-cart. One reporter, writing for the business blog *Quartz*, compared the new, on-demand jobs enabled by smartphones to menial jobs he saw growing up in Mumbai, India: doing laundry, delivering lunches, and driving cars for the wealthy. You didn't need technology to make these jobs happen; you just needed a large pool of laborers willing to work for pennies, motivated chiefly by poverty and inequality.

Technological inequality is a term many have cited as a big cause of the hollowing-out of the middle class in America and other devel-oped nations. "Tech inequality gets going in the 1990s, and stacks on top of financial sector inequality," said Tyler Cowen, a professor of economics at George Mason University who believes inequality is inherent in the nature of technology jobs. "The trend will continue. Major tech companies don't employ many people. They do indirectly create jobs, but they tend to be low-paying service jobs. The cognitive requirements for working with tech are so high, it's antiegalitarian. You get all of these effects together, and the digital economy is an un-equal economy."

Silicon Valley may be booming, but that doesn't help the hun-dreds living in homeless camps just miles away. Research by the Brookings Institution found that San Francisco saw a vast increase in inequality from 2007 to 2012, more than any other US city. Occa-sionally this issue gets pushed to the forefront, as it did in 2013, when groups of protesters in San Francisco blocked the private buses that shuttled workers at technology firms to their jobs in Silicon Valley. The objective of the protests was ostensibly around these so-called luxury buses abusing taxpayer-funded infrastructure, but its root was the increasing divide in that city between the tech haves and the non-tech have-nots.

*N*one *of this is lost* on Jacques Panis, the swaggering president of Shinola. Prior to Shinola, Panis ran an online virtual community for children and a visual effects company that Kartsotis later purchased. I asked him how running an analog company differed from running a digital one, and Panis held up his wrist. "The product," he said, looking at a Runwell watch he wore. "It's not just living on some server in God-knows-where. This isn't lines of code. I wake up every morning, look at this thing, and it says, 'Seven thirty, go to work!' It's such a much more rewarding business to be in, because I know the people behind it at every step of the process."

Yes, Shinola's watches were out of reach for most Detroiters, but so were Cadillacs and Corvettes. The company was an anomaly in the way it operated, a "freak" created and funded by a visionary multimillionaire who didn't have to answer to investors or the short-term demands of the stock market. What mattered was the work and increasing demand to the point where Shinola produced enough watches in Detroit that it could attract a domestic supply chain, which could in turn employ more Americans. "The fundamental idea here was creating jobs in the US and building products that we can make here today with muscle, might, blood, sweat, and tears," Panis said. "Our hope is in time there is a guy who says, 'Hey, I can make your cases; I can make your crowns and buckles.'"

It was a heartwarming story, but was it good business? If the goal is to maximize profits, then saving money on labor, the highest cost in most businesses, would translate into higher profit. If workers in Thailand, or even machines, could assemble Shinola's watches cheaper than can the people in Detroit, then why not go that route? If you stripped away any supposed emotional epiphanies about the American worker by Kartsotis, and looked at Shinola from a purely rational basis as an investor, wouldn't an automated, offshore watch company be more profitable?

In the short term, yes, that would likely be the case, but the continued and growing employment of Shinola's American workforce is the core of Kartsotis's investment thesis for the company, which is based on two factors.

The first is emotional. Although Shinola watches are priced as entry-level luxury watches—costing more than a $200 Fossil watch, but less than a $3,000 Rolex—the key selling point for the brand isn't so much its design, heritage, or price, but the story behind it. And that story is about Shinola's workers making watches in Detroit. Move that production overseas or automate it, and a Shinola watch's worth plummets.

It is this story that people are buying and strapping to their wrist, and it is one that the luxury watch industry only discovered after being disrupted by technology. The Swiss traditionally dominated watchmaking, but when quartz movements were introduced in the 1970s, and later digital watches, Swiss watchmaking took a dive, as competitors in Japan produced cheaper, more accurate watches. The Swiss struggled for nearly thirty years, but by 2008, Switzerland was once again the leading exporter of watches by dollar value. The reason for this, according to Ryan Raffaelli, a professor at Harvard Business School who wrote a comprehensive study on the Swiss watchmakers' revenge, is something called technology reemergence. "Technology, as it becomes more sophisticated and mature, eventually gets displaced by newer or more effective technology," Raffaelli told me, describing the traditional path of creative destruction. "But there are these odd circumstances where there's an alternative path, where these dead technologies get repositioned for new life."

For the Swiss, this meant reframing their technology. First, they created the Swatch brand, an affordable, fashionable line of watches, which showed they could compete on emotion rather than precision. Second came the renewed notion of craftsmanship, which accrued an elevated luxury status for such brands as Patek Philippe. "You're not just experiencing this technology and its value, but the craft that's been passed down for generations, and you can connect to human hands that formed and built it in the first place." That knowledge is the luxury a watch buyer who pays $50,000 or $100,000 for a Patek Philippe timepiece is purchasing, which vastly outweighs any material worth of the watch's components or the accuracy of its timekeeping. If you want accurate time, look at your phone.

Raffaelli told me that the emotional brand Shinola built around its workers protects its watches from pricing concerns in a way that is different from Swatch or Fossil, where the worker is anonymous. Shinola's workers aren't just the people who assemble the product; they *are* the product, and that prevents the product from becoming a commodity good whose price is the main factor in its success.

The other half of Kartsotis's investment thesis goes back to the notion of reskilling at industrial scale. Shinola's leather watch-strap production was created out of necessity, when the company's supplier couldn't keep up with demand. Once Vega and others were training Shinola's workers in the craft of fine leather manufacturing, it installed the capacity to create other leather products. So, Shinola started making leather bags, wallets, and other accessories at its factory with the same workers and machines. Kartsotis told me that this has put the company a hop, skip, and jump away from making such products as shoes, which could create many more jobs than watches. Each new skill taught to Shinola workers would lead to more products, which would lead to more new skills, and more products beyond that. This is how Kartsotis envisioned the future of Shinola: growing from a small niche into something much more mass market.

When I spoke with Kartsotis by phone (he didn't want to be quoted directly, so I'm paraphrasing his comments), he told me the Shinola factory was built to assemble half a million watches a year, a number that it would soon reach. That could probably double to a million watches annually, which could lead to a couple of thousand jobs in Detroit, in addition to other product lines. This would be a drop in the bucket, employment-wise, compared to an auto parts plant running at three shifts, but for Kartsotis, who wants to take Shinola public on the stock market, it would prove out his investment in the brand as an emotional idea rather than a specific product (in 2014, the company reported revenues of $60 million, but was not yet profitable). That success would be important, not just for Kartsotis, Shinola's shareholders, and its employees, but for the larger population of entrepreneurs and investors in Detroit and beyond, who could look at Shinola's success as a model of an analog business to emulate.

Analog industries, such as manufacturing and retail, need profitable success stories, because one of the biggest narratives of the digital economy is that it is the best place to get rich. The huge stock market valuations of such companies as Facebook and Amazon feed this storyline, as does the news of startups with absolutely zero revenue being acquired for billions of dollars. Who wouldn't have loved to invest early in something like Instagram and retire spectacularly wealthy? Thanks to its ability to scale rapidly, a digital business can multiply its value ten, twenty, or a hundred times in just a few years. That is almost impossible with an analog business, whose growth is hemmed in by the laws of the physical world. People need to be hired; warehouses need to be built; supply chains have to be figured out. All of this takes time and money, which is why even low double-digit growth is seen as really good for an analog business.

There's a flip side to this, of course. Because it is a winner-takes-all industry, investing in a digital business is much riskier than investing in an analog one. Venture capitalists might seem to make hoards of money, but most of their success comes from investing in a large pool of deals, hoping that one will hit it big and return a hundred times the investment, offsetting the losses from the other startups that failed to make any money. In contrast, someone who invests in analog businesses, such as a private equity firm, will invest in a smaller pool of business and use the funds to grow them over time.

Investing in analog businesses is a long, patient game of hitting singles and doubles to rack up runs. Investing in digital ones is waiting for the long ball. You might miss ninety-nine pitches, but none of that matters if you whack the hundredth out of the park. And the truth is that most technology startups will fail and be left with nothing to show at the end of the day, because they hold few concrete assets. A housing development that goes bust still has land that can be sold or held until it appreciates in value. If you invest in a technology company and it goes under, the most you can hope is to recover some of its office furniture. There is no long-term value in code. It is not an asset.

This matters in the greater scheme of things, because the notion that investing in technology is the way forward for households, institutions, and communities is more like a compelling sideshow than the story of the real economy. Multibillion-dollar valuations for such companies as WhatsApp and Uber distort reality. They are the exception rather than the rule. "New technologies have yielded great headlines, but modest economic results," wrote Nobel Prize–winning economist and columnist Paul Krugman in the *New York Times* last year. "You see, writing and talking breathlessly about how technology changes everything might seem harmless, but, in practice, it acts as a distraction from more mundane issues—and an excuse for handling those issues badly."

There is also the question of what type of investment best serves a community. In Detroit, the needs of the community and its underlying assets are overwhelmingly analog. "Forty-seven percent of adults in Detroit are functionally illiterate," said Gary Sands, a former Detroit city planner and retired professor at Wayne State University, as we ate lunch next to the Shinola store. "Compuware [a Detroit software company], Twitter, and Google aren't gonna do these folks any good. They're an analog population."

The problem is that analog jobs aren't sexy in the way tech jobs are to politicians, investors and philanthropists, and the media.

"When Twitter put one person in an office downtown, it made the same headlines as Shinola opening its factory here," said Kyle Polk, a real estate developer with the local firm Town Partners and a consultant to the Detroit Future City project, which gathers and analyzes the city's economic data. Polk grew up in the city of Detroit, and returned after working in investment banking and for the Federal Reserve Bank of New York. He was raising his family in his grandmother's old house, and has no illusion about the reality of his community's needs and the role that digital technology companies can play in improving it.

"You think you'll create these live-work-play communities by putting in a Whole Foods," Polk said, referencing various urban revitalization schemes and developments being floated as solutions to Detroit's problems. "But these people who don't have jobs and are hungry won't work-play-live. Motherfuckers will rob you in the Whole Foods parking lot! You've only made them more frustrated. . . .

The majority of people outside the labor force in Detroit don't have a college degree," Polk said. "If you want to bring jobs into the community, why the fuck would you bring in college jobs? Analog isn't a growing trend, it's smart business. If you had a choice to bring in [a distribution warehouse] and Yahoo! why would you not take the one benefiting the labor pool?"

Polk saw Shinola as a more suitable model for the city: niche manufacturing and services that had the capability to scale, and built off the human and physical assets already present in Detroit, which made it attractive and competitive to investors. Polk's firm had recently purchased several buildings around Shinola's factory (which is housed in the city's largest design college), and was developing a craft district of small and medium-size factories making labor-intensive, high-end consumer goods. There are a number of new manufacturing companies in Detroit along these lines, making everything from pet accessories to denim jeans, bicycles, beard balm, kitchen counters, frozen food, and high-end furniture. Some of these are tiny—really no more than a few people indulging a hobby—but others are major businesses that now occupy huge facilities and employ hundreds. Manufacturing remains Detroit's greatest asset.

"Analog can be sexy," Polk said. "And if you want more people in that space, it has to be sexy."

Shinola had staked its entire existence on this notion. In the months after I visited its factory, it launched a production line for electrical cords and lighting accessories, and announced it would begin manufacturing audio equipment, including turntables and headphones. This coincided with a new partnership between Shinola and Third Man Records, which included a Third Man store opening in Detroit next to Shinola's own shop, with a brand-new Third Man vinyl record–pressing plant set up behind it. It was a fitting return home for Jack White and Ben Blackwell, whose music careers began in the very neighborhood.

When I spoke with Kartsotis on the phone, he estimated the Third Man plant could press as many as 10 million albums a year, nearly as many as United Record Pressing did. All of this would create more jobs, more salaries, and put more money into the hands of Detroiters. The city may have been bankrupt, plagued by crime, inequality, poverty, and unemployment, but in this little analog corner of Detroit, business was booming.

8

The Revenge of School

There are few easy solutions to digital's disruption of the jobs market. Some have proposed a mandatory minimum income, while others talk about the need for greater government investment in infrastructure, and subsidies for labor-intensive industries such as energy and manufacturing. The one solution that is almost unanimously accepted is the need for better education. From world leaders to economists, tech-industry gurus, and eager young teachers, creating the future of education is a mission that resonates almost universally.

How education's future will be delivered to students is another story altogether, and it was one that Christopher Federico and Karen Wolf were tackling one arctic February morning in front of a classroom of teachers at the University of Toronto's Rotman School of Management. Federico and Wolf are both full-time teachers themselves: he teaches problem-based learning at the gifted high school run by the University of Toronto and she teaches English at North Toronto, a public high school. The twenty-odd teachers before them came from a variety of backgrounds, ranging from kindergarten educators to community college professors, and were here for a two-day course the university offered in Integrative Thinking for educators. Integrative thinking is a methodology for complex problem solving used by management consultants, which the Rotman School taught to its MBA students. A few years back, Rotman began offering short courses to educators in integrative thinking, so they could teach these methods to their own students and build problem-solving skills into the curriculum.

Wolf and Federico began the lesson by asking everyone to quickly draw a portrait of the person next to them. After a minute, everyone revealed their portraits, and laughter rippled through the room. My portrait of Jeff, a fifth grade teacher from the nearby city of Guelph, resembled an anorexic robot, though to be fair, he made me into a flat-topped Frankenstein. "This will only work if we go a bit outside our comfort zone," Federico said, explaining the point of the exercise. "Focus less on what you draw and more on *how*. This is not an algorithm. We are looking for a way of approaching the world in a manner that's more reductive than breaking problems down in charts to spit out the one right answer." The goal was not to make compromises between sets of available choices, or to think "out of the box" with random ideas, but to explore and examine the available options, and use the best of those to create innovative new solutions.

Federico drew a line down the center of a whiteboard. "What is the future vision of what school looks like?" he asked the class.

This was not a rhetorical question, but the problem these teachers would tackle today, first by comparing and evaluating two apparently contrasting models of education and later using the data to create a new approach for schools. One model was the brick-and-mortar school, the analog bedrock of teaching that exists the world over, and the place where all these teachers worked. The other model was the online-only, virtual school, a digital alternative that Federico said seemed to be the way of the future.

Wolf then asked the teachers to list only positive attributes of each model, as they would be making something called a pro-pro chart. "Nothing negative," she said, "only pro here." Teachers called out ideas for each: Online-only schools could connect students to teachers anywhere and anytime. They could be more cost-effective, and nearly every aspect of the experience was customizable to the individual needs of students. Teachers could even work from home, in their pajamas . . . a comment that elicited whoops of approval.

In terms of advantages, brick-and-mortar schools were situated in a particular community, and students could form deep social bonds with teachers and peers there, what Federico called the "hidden curriculum" of socialization. Educators in traditional schools got a job, a sense of belonging and purpose, and the reward of seeing students learn in front of their eyes.

Presented here as brightly as possible, these two models for education showed a harmonious, positive future for schooling, whether in-person or online. But out in the real world, the future of school and the role of digital technology in it have become one of the most hotly debated issues in the public interest. Education, especially in the United States, is often referred to as "broken" and "failing." In global assessments and test scores, American students perform meagerly, far worse than those in other wealthy nations, and often less than some developing nations. Education reform has become *the* great cause in America, and various stakeholders champion a host of solutions to save it.

Few industries have embraced the desire for radical, transformative change in education with the zeal, enthusiasm, and commitment of the digital technology industry. This makes sense for two key reasons. First, education is a prized pig, ready to be roasted and devoured by digital disruption. Today, total spending on education technology remains low, around 5 percent of total education budgets in the United States, and less than 2 percent globally, but worldwide spending on K–12 classroom hardware technology alone is expected to reach $19 billion by 2019, and 2014 saw more than a 50 percent increase in venture capital funding for education technology companies. That's a lucrative market to tap into.

Second, the high-tech world is fueled by education. Its businesses are created by highly educated individuals, often at universities, and many of the products and services it sells appeal to an educated population. Education has become the pet cause of digital's business leaders. Bill and Melinda Gates, Mark Zuckerberg, and the venture capitalist Jim Breyer are among the top supporters of education philanthropy, funding everything from university scholarships and research grants to experiments in school reform stretching from inner-city Newark to remote African villages.

Underlying this is the belief that digital technology can transform education in the same way it transformed business, media, and communications. What emerges is a vibrant, multibillion-dollar market in education technology (ed tech, as it's commonly known) that promises nothing less than a radical rethinking of education. Here is where the utopianism and manifest destiny of Silicon Valley meet your child's elementary school, and where pedagogy and philosophy intersect with politics and business. Attend a presentation of an ed tech company, watch a TED talk about education, or listen to a school superintendent talk breathlessly about the new virtual-reality goggles she just bought for your kid's school, and the future is bright indeed.

It is a future where every child has the ability to learn at their own pace, in the most stimulating way possible, from wherever and whenever suits them best, at a lower cost but with greater accountability and results. It is a future where school will be dynamic, where teachers will truly be able to unleash their creative potential, where inner-city teens will have the same advantages as those in wealthy suburbs, and the greatest university in the world will not be some ivy-covered campus, but anywhere your device gets a signal. The old, ineffective system of sitting in rows of desks, listening to a teacher regurgitate information from the pages of books, will be turned on its head. We won't need their education. We won't need their thought control. The walls will come down, and a bright new future will emerge.

That's the promise, at least.

The reality of digital education technology, which has attempted to realize this future for much of the past thirty years, is that of troubled students who have shown tremendous promise but consistently gets D's on their report card. It is a cautious tale of what happens when schools, communities, and educators place their blind faith in digital innovation while ignoring the proven evidence and research around the benefits of analog education. Time and again, analog schools and teachers have proven not only better at teaching students, but as I witnessed at several schools around Toronto, they can actually present more innovative solutions for education's future.

———— * ————

*T*his *story is not unique* to the digital era. The inventors, manufacturers, and evangelists for radios, mail-order correspondence courses, television, VCRs, and even the printing press all made grandiose predictions that their technology would either transform traditional schools or eliminate them entirely. Thomas Edison himself proclaimed that books and teachers would soon disappear from classrooms, because students would learn through the motion pictures he helped invent. The birth of the digital computer just added more claims to this long history. The latest educational software or device is always unveiled with the same breathless belief in technology's potential to disrupt school.

"There's been this pattern of hype claims of the transformation of schooling, leading to academic studies that bred dissolution, and ultimately blame on teachers," said Larry Cuban, a professor of education at Stanford University. "This pattern goes back well over a century." Cuban, who lives and works in the heart of Silicon Valley, began as a hopeful evangelist for education technology, but slowly turned into one of ed tech's most prominent skeptics after witnessing, time and again, the failure of ed tech to deliver on its promises. He calls it the hype cycle. "There is this pattern of extreme claims for transformation, and then a kind of bumpy landing and disappointment."

Why does this happen, over and over again, without the technology industry, the educational institutions, and other stakeholders learning from their mistakes? It is not as though the evidence is lacking, or industry leaders lack the ability to learn from mistakes. Rather, Cuban attributes the persistence of ed tech's hype cycle to deeply held values around technology and innovation. "In this culture, like other developed cultures, technology is seen as an unadulterated good," he said. "The presumption is that the technology will improve one's life, in whatever it is." Education's stakeholders are often blinded by this presumption of technological progress as the ultimate good and cannot look at its actual performance critically.

"The skepticism that one would ordinarily raise about inflated claims comes pretty late in the process when it comes to anything technologically innovative," Cuban said. "Any skepticism about decisions that buy and distribute electronic devices [for schools] tend to be rushed into very quickly. And a lot of money gets spent. Why? It doesn't matter what the research studies say, or what any doubters in this field say. Anyone who doubts is called a Luddite. All of that is suspended because of the rush to get [the new technology] into class-rooms. When it comes to schools, they are heavily dependent on vot-ers and taxpayers. They've historically been seen as behind the curve of the private sector and modernizing. When it comes to technology, it's very important for school boards and trustees to say 'We're at the cutting edge. We bought these iPads for kindergartners!' Teachers are rarely involved in those decisions, and these devices show up at the classroom door."

Cuban cites three reasons that policymakers typically use to jus-tify the purchase of new technology for schools. First, the technology will improve student achievement and marks. Second, the technology will change traditional teaching to nontraditional teaching. Third, the technology will better prepare students for the modern workplace. At best, Cuban says, there is contradictory evidence for the third rea-son, little for the second, and none for the first.

---

*T*o understand why education technology fails so frequently, it's im-portant to start at the beginning of our learning life, a period known in the field as early childhood education (ECE), which covers daycare, preschool, and kindergarten. While many activities during this time may seem like a lot of aimless playing, naps, colds, and di-aper changes, it is actually the most crucial educational experience of our lives, because it provides the foundation for all our learning that follows. Young children learn about the world through physical senses: grabbing and touching, smelling and hearing, seeing, licking. The widely held recommendations by pediatricians the world over to avoid exposing children under age two to screens is not out of concern that the content on those screens will damage their brain, but for fear that they will replace more valuable, sensory activities, such as putting their hands through a box of sand, or eating a tub of Play-Doh.

"The big organizing ideas around our formations of relationships are that physical experience," said Diane Levin, professor of early childhood education at Wheelock College in Massachusetts. "ECE theorists say that's the foundation for both learning [and] social, emotional, and cognitive development." Levin used my own daughter's experience at daycare that day as an example. At the time, she was one and a half, and was finger painting in her class. That activity not only involved her ability to create an image on paper, Levin said, but the sensory feeling of the wet paint running down her arm, the visual learning of the colors mixing as she moved her fingers around the paint, the spatial learning when she moved her arm off the paper and the paint dripped onto the floor, and the social learning when she flung paint at another kid and they cried, and the teacher told her why that wasn't cool and why she had to apologize. Finger painting was a full-body, full-mind experience. Compare that to numerous finger painting apps available for a tablet, and the sensory learning experience is reduced down to the tips of her fingers dragging across a small glass surface, without texture, smell, taste, or other physical and social consequences. "When you're pushing buttons, it's an abbreviation of all of that," Levin said. "You're just not getting it."

Even the best educational computer programs and games, devised with the help of the best educators, contain a tiny fraction of the outcomes of a single child equipped with a crayon and paper. A child's limitless imagination can only do what the computer allows them to, and no more. The best toys, by contrast, are really 10 percent toy and 90 percent child: paint, cardboard, sand. The kid's brain does the heavy lifting, and in the process it learns.

A big component of early childhood education is play-based learning, which is the guided use of unstructured play. Children need to explore the limits of acceptable behavior through play, and the physical and social consequences of those actions. Dorothy Singer, a retired professor of child psychology at Yale and one of the leading researchers on play-based learning, told me a story about observing two young children building a house out of blocks some years back. Four other kids came and decided to knock it over, and when they did, the two original builders ran to the teacher, crying. But a few minutes later, those two kids began negotiating with the demolition crew on how the house should be rebuilt: how many doors, how big it should be, who would do what job in the construction. "As the narrative moves along when children play, it affects their mind and they learn the rules of society: how to make something happen, how to live with each other, how to avoid confrontation," Singer said. "You don't see that when a child plays on a computer."

All of this is necessary, even as children inevitably grow up to use computers in their later schooling or work. Education is a lifelong building process that starts with a foundation of very basic skills and increases, year after year, in its complexity and abstractness. When I am typing these words on my laptop, I am using spatial and social reasoning skills that I learned as a three-year-old playing with LEGO bricks. "With parents, there's the belief that we live in a digital age, and it's a good idea to give them the technology early," said Jeff Johnson, an early-childhood-learning author and partner in the business Ooey Gooey, which makes play sand and other learning toys for preschoolers. "But just because they'll use a piece of technology when they grow up, doesn't mean we have to give it to them now. We don't give a kid a chainsaw at two and say, 'Here you go!'"

Part of the problem is the confusion between the enjoyment kids get from technology and its educational benefit. It may seem like an act of sheer, adult brilliance when a one-year-old can unlock your phone, open a music app, and select a song, but it isn't. It is just a fancier version of the talking doll with a string in its back, or other toys with batteries and bright buttons that primates can also play with. That doesn't stop companies from devising and selling dolls, cribs, and potty seats that work with iPads, but we have to ask whether we are actually teaching kids anything with these technologies or just entertaining them.

A quick caveat here: I am not damning the wholesale use of digital technology in education. Digital technology can make education more effective when used appropriately. Schools run more efficiently thanks to the use of computer systems, which manage everything from report cards to budgets. Teachers and students can use computers to research, write, create, evaluate, correct, and manage their own educational environment. Academics from around the world can coauthor studies, evaluating far more data, far more quickly, while kids with special needs (autism, ADHD, dyslexia) have been shown to respond effectively to digital learning tools and environments in many cases. The criticism around educational technology also does not apply to the teaching of computer technology itself. Computer programming, coding camps, maker clubs, and robotics competitions are all valuable and necessary for teaching the knowledge and skills of those who wish to learn about digital technology. These are growing and increasingly important fields.

But including a mandatory course in computer programming for students is a very different thing from what the majority of ed tech evangelists hope to achieve, which is the integration of digital technology across all schools and subjects. It is rooted in the idea Cuban spoke to—that technology equals progress—and the more it can be woven throughout the school experience, the better off students will be. At its most optimistic and dangerous, education technology arrives as the transformative panacea that will fix education and leaves a trail of disappointment and failure in its wake. The evidence for this just keeps on piling up. Study after study seems to confirm how the implementation of educational technology produces little net benefit to student performance, and in many cases, actually makes things worse. The examples cited here, which represent just a fraction of the existing and ongoing research into this, show the various ways educational technology falls short.

One of the big beliefs in the ed tech movement is the need to bridge the so-called digital divide between those who have access to computers and those who don't. The theory holds that wealthy students with computers do better than poorer students who don't have access to those computers. Increase access to computers and the Internet, in schools and at home, the thinking goes, and watch inequality fall. This is a project politicians, parents, school administrators, philanthropists, and the media have taken to with great gusto, because it presents such a beautifully simple solution to the intractable problem of educational inequality.

A 2010 study by Duke University tested this theory out by looking at North Carolina public school students who were given free laptops, and what it found was the diametric opposite. "The introduction of home computer technology is associated with modest but statistically significant and persistent negative impacts on student math and reading test scores," the study's authors wrote. "Further evidence suggests that providing universal access to home computers and high-speed Internet access would broaden, rather than narrow, math and reading achievement gaps. . . . For school administrators interested in maximizing achievement test scores, or reducing racial and socioeconomic disparities in test scores, all evidence suggests that a program of broadening home computer access would be counterproductive."

The same logic of bridging the digital divide was behind the wildly ambitious One Laptop per Child (OLPC) nonprofit, spearheaded by MIT Media Lab founder Nicholas Negroponte and set up in 2005 with the backing of a vast coalition of philanthropists and technology companies. OLPC's goal was to produce and distribute rugged, inexpensive, Internet-enabled laptops to the world's poor with innovative features such as solar panels and hand cranks. OLPC successfully created several devices that met this goal, but in every other respect, OLPC was a colossal failure that typifies the hubris of tech-centric educational utopianism.

From the outset, education ministers and development professionals pointed out that what children in rural Pakistan or Rwanda needed most were safe schools, clean drinking water, and trained teachers . . . not computers. OLPC nevertheless pressed ahead, and sold nearly 3 million of its custom laptops to schools around the world. Negroponte loved telling the story about OLPC distributing tablet computers to remote villages in Ethiopia with no schools so children could teach themselves. Then the evidence emerged. Across continents and countries, from Peru and Uruguay to Nepal, well-funded academic studies demonstrated no gain in academic achievement for OLPC students when compared with those who didn't participate in the program. The evidence mirrored other laptop and computer handout programs in such countries as Israel and Romania, where the introduction of computers also did nothing to improve learning. Last year, a report from the Organization of Economic Co-operation and Development concluded that "students who use computers very frequently at school do a lot worse in most learning outcomes," and technology did nothing to improve scores across subjects, and less to bridge gaps between rich and poor students. In 2014 One Laptop per Child closed its Boston headquarters and drastically cut down on staff and new programs.

OLPC's great mistake was presuming the universal importance of a shiny imported technology in spite of the recommendations of people closer to the problem at hand. This problem is not confined to international development. In Detroit, I spoke with Amanda Rosman, the cofounder and executive director of the Boggs School, an innovative charter school in a poor neighborhood. She believed that the imposition of technology as a solution for underperforming, urban schools is rooted in many of the same racist attitudes behind other misguided urban education policies, and is widely abused. Charter schools in Detroit, which receive state funds based on enrollment numbers, regularly use technology to beef up student registration. "If you drive down the street before Count Day [the cutoff date for charter enrollment], there are banners hanging by these schools saying, 'Every kid receives their own iPad!'" Rosman said. "It's an incentive. A prize."

The amazing thing about all this is how the people most responsible for education policy fail to learn from their mistakes. In 2001, the Los Angeles Unified School District spent $50 million on a computer system called the Waterford Early Reading Program, created by the education publisher Pearson to improve language instruction in kindergarten and first grade. Shortly after, research by the district discovered zero or negative reading improvement for students who used the program. When Waterford was abandoned in 2005, the school board's president at the time, Jose Huizar, told the *Los Angeles Times*, "How could anyone continue to argue that it's working when it's not? It's underutilized and ineffective."

Nine years later, the very same LA school board announced a plan to put an iPad into the hands of all its 650,000 students. The iPads were loaded with Pearson educational software and coupled with a big push to improve Internet access at LA schools, all at a total cost of $1.3 billion, one of the largest single ed tech investments worldwide. Shortly after the first batch of iPads was distributed, the rushed, ill-conceived folly of the entire enterprise became apparent.

The iPads had no keyboards, which made them useless for students to do homework on, and the software that was supposed to prevent students from using the iPads for games and social media was easily hacked. The iPads frequently malfunctioned, were lost or stolen, and the software was inadequate for learning and assessment. To top it all off, the FBI launched an investigation into whether Apple and Pearson had received preferential status as vendors over other potentially less expensive competitors. Barely one year after launch, LA's iPad program was canceled and the city's school superintendent resigned in disgrace.

From failed laptop implementations in Hoboken, New Jersey, to the tales of cracked screens, melted chargers, and tremendous financial losses for News Corp's Amplify tablet program, time and again the mass "airdrop" of new ed tech devices into schools has fallen flat. But the attraction of politicians and policymakers to ed tech's charms remains irresistible for a number of reasons.

One is political. Announcing you are going to give out iPads to every child in a community appears as a bold, clear signal that you are investing in the future and aligning your schools with the biggest, most innovative company in the world. It steers clear of any sticky issues with powerful teachers' unions, and provides for great photo opportunities and news stories.

Another big reason behind the eager embrace of technology, especially with public school boards, is the promise of savings. With the help of digital technology such as computer-assessed standardized materials and tests, a school board can theoretically achieve economies of scale. And the hope is that once the learning becomes effective through devices, a school board should need fewer highly paid teachers and professors, who can be replaced with facilitators and teaching assistants hired to aid in the digital learning and exams, while the computer does the heavy lifting.

The temptation for eventual savings is powerful, but it underscores that the implementation of technology in schools carries a financial burden. Not just the initial capital cost of acquiring the technology, but continual expenses to maintain, repair, replace, and update it. A school gymnasium can last decades, a good textbook sometimes fifteen years or more. Some of the desks at my university were damn near a hundred years old. But any digital technology, no matter how well designed it is, becomes obsolete in just a few years, and inevitably stops working. My only memory of school computers was of dusty relics in the corner that didn't even turn on.

Dollars spent on digital education technology are dollars that cannot be spent on teachers, building maintenance, or textbooks. It is money that has been pulled from programs in art, sports, music, and drama. Even though the research shows one of the greatest factors in reading improvements in students is the presence of school libraries, the number of libraries across school boards in the United States has declined dramatically. The logic behind this is often that libraries are pointless in the age of Google and eBooks, and that money would be better spent buying tablets or drones.

In his riveting book *The Flickering Mind*, Todd Oppenheimer chronicled the failure of various education technology initiatives in America, and the real cost they imposed on the schools that adopted them: "In debates about the importance of classroom basics, the technologists often argue that they aren't trying to displace solid fundamentals. Tech isn't meant to be a replacement, they say, it's a supplement," Oppenheimer wrote. "The line is hollow . . . an 'e-lusion.' Trying to fully support technology initiatives is extremely costly. Beyond the financial expense, there are the demands that computers make on a school's time and energy. . . . these are not flexible resources; every community can only offer a fixed amount of each one, and any amount devoted to technology leaves less available for other practices. So when technologists argue that tech is only meant as a supplement, they're either fools or liars."

One winter morning, I sat in on a fifth grade class at the Jackman School, an elementary laboratory school that is part of the University of Toronto's faculty of education, where new theories and technologies are tested in real classrooms. Benjamin Peebles, the teacher, had gathered his students around the room while three of them read aloud from press coverage they had printed off that morning, which the class then discussed. I asked Peebles why he made the kids read these stories from paper when the school provided laptops and iPads for all its students. Peebles told me his students actually chose to use paper themselves for the assignment, even after he had given them the option of reading them off a device. Considering these were all stories printed off websites, I asked the students why they made this choice.

"It's sort of more comfortable to bring in a sheet of paper with words on it," one girl, wearing a Katy Perry T-shirt, said. "It just feels right."

Another said it was impossible to highlight and underline on a tablet, while a boisterous young boy told me, "It gives me a sense of closure to have something in my hand." Soon, I was surrounded by nine-year-olds shouting out their preferences for paper: It was easier to share. It felt more real than an iPad. It felt better to touch a page than slide your finger on glass. It was easier to concentrate. One girl, referring to writing on paper, told me she would rather be limited by the space on a page than have her creativity limited by a computer program.

These kids' preferences for paper were increasingly supported by research showing how a majority of students prefer learning with paper over digital formats. The reasons students in these studies frequently cite are identical to the ones the kids at the Jackman School mentioned: paper books are easier to navigate and personalize with notes and marks, more reliable (a textbook will not malfunction, freeze, or delete its text), less costly and more versatile (you can share books, borrow them from a library, buy them used, and sell them), and better for learning. Reading print books, taking notes on paper, even banning laptops in classes . . . all have been shown, and proven, to actually improve student performance and information retention. Why wouldn't you stick with a book?

One academic behind several of these studies is Dr. Joanne Mc-Neish, an associate professor of marketing at Ryerson University in Toronto. McNeish grew interested in student reading habits a few years back when she noticed a disconnect between her undergraduate students, who mostly seemed to be reading and writing on paper, and her colleagues, who consistently said, "These kids all just want e-books." When she studied students in Canada and Israel, McNeish discovered something interesting that linked them all: students overwhelmingly prefer paper not out of any sense of nostalgia or a resistance to new technology, but because paper learning materials simply work better. "It's a lot of work to use these e-learning systems. And a lot easier to just learn from a textbook," McNeish said. "These kids are skilled with technology for entertainment, but they are not so skilled at technology for learning."

McNeish felt the push for digital textbooks and other electronic education technologies was not driven by students, but by older faculty and administrators, who made faulty assumptions about their young students. "The fatal flaw about baby boomers is that they're really afraid of not being hip or cool," McNeish, herself a baby boomer, told me. Students and schools needed stability as much as they needed change, she felt. If everyone was disrupting and reinventing everything all the time, how would anyone move forward?

———⋅———

Digital technology may grab all the headlines around educational innovation, but in many instances analog ideas and tools are vastly more promising for effecting change. "All of the most important ideas in education reform have nothing to do with ed tech," said David Noah, the managing director of the Success Academy charter schools in New York City. "They involve teachers collaborating more, longer school days, state-funded pre-K, money for after-school programs . . . these are solutions that people know will work, if done right. These are not technology driven reforms. Technology is at the periphery of reforms."

Noah characterizes the majority of educational technology tools as digital imitations of existing analog ones. An electronic textbook has the same content as the paper edition, and all its other features (live discussions, hypertext links, embedded videos) may look cool, but does nothing to improve learning and often just provides

distractions and technical problems. "The 'tech revolution' in education is a bunch of people taking very rote versions of stuff, and making digital versions of it," Noah said. The most commonly used math
software presents the same quizzes as standard math textbooks, but
on a screen, with quicker marking. "It mimics not teaching, but the
worst idea of what teaching is: a teacher saying, 'Do problems one to
twenty,' and then spacing out. When people talk about great teaching, that's not what they're talking about."

A big reason why ed tech so often fails to deliver lasting educational innovation is that it is often created and implemented with minimal input from teachers and students, the end users of educational
technology. Teachers are a broad, diverse group who differ greatly in
each community and school. Ed tech solutions are most often created
with the hope that they can be sold to a broad spectrum of schools
and classrooms. They tend to be one-size-fits-all solutions, and while
these companies may include a founder who is a teacher, they are
often not teacher led. But there is another, more insidious notion underlying much of ed tech's troubled approach to innovation, which is
that the teacher is an obstacle to be overcome.

Teachers, especially in public schools, have been characterized as
lazy, entitled, unionized dinosaurs who are wholly resistant to progress. Many of the most popular ideas in education reform today have
this bias running through them, from the Common Core curriculum
and teacher compensation linked to test scores to the Thiel Fellow-
ship, an antischolarship from PayPal cofounder Peter Thiel that pays
promising university students to drop out and pursue a startup instead. When ed tech fails, the blame is often placed on the teachers
who apparently didn't adopt it correctly or with enough enthusiasm.
This assumption is as arrogantly false as handing out laptops to poor
children and expecting their lives to change. It ignores reality.

The reality is that teachers know what works best in their classrooms, and they know what they need to improve the way they teach
their students. It took Shawn Collins years to realize this. Collins is
a former teacher from Massachusetts who is now the director of business development at PolyVision, a division of Steelcase that makes
whiteboards. In the 1990s, he joined a company called Microtouch,

which was one of the early manufacturers of so-called smart boards, which are digital whiteboards that link up to a computer and allow their users to "write" on the screen, while also accessing graphics and other files from a computer. At the time, smart boards were presented as the greatest innovation in education technology.

"I spent eight years of my life telling people you didn't need a whiteboard in the classroom anymore," Collins said. "Teachers rebelled against that. We thought it was a training issue, and that if they had the right training they wouldn't use the whiteboard. But it wasn't training . . . it was a use issue. Teachers just wanted to pick up a marker and write on a board, and you can't do this with technology." These teachers weren't stubborn or resistant to technology. They were open to new tools and ideas. But they wanted a technology that worked for their needs.

Eventually Collins began to listen, and PolyVision created a new product that flanked a smaller smart board with up to twenty-four feet of standard whiteboard. This provided the best of both worlds, because teachers could now introduce new visual material on this central screen, but could also write a huge amount on the whiteboard. Subsequent studies have demonstrated having lots of whiteboard space in a class helps students retain information through something called information persistence. If information is visible for days or weeks at a time, it soaks into the brain a lot more readily than something that appears on a screen for a few seconds.

Whiteboards and such products as IdeaPaint (a popular whiteboard paint) are two of the more frequently requested items from teachers on the educational crowdfunding philanthropy website Donors Choose, which has seen an increase in overall requests for analog classroom tools since 2014, and a similar decrease in requests for digital technology. The largest category for teacher requests on the site are for books, and the most requested item is a black marker for whiteboards. "The origin of the site was very much teachers that felt analog/physical resources were lacking in classes," said Charles Best, the CEO of Donors Choose, who founded the nonprofit in 2001 while teaching history at a public high school in the Bronx. While he found that "money [was] always made available for the flavor of the month technology," Best and his colleagues regularly lacked supplies as basic as pens, paper, and up-to-date textbooks.

———————◆———————

Where the real lasting innovation in education lies is not in hardware or software, but in new approaches to teaching that shapes how students learn. One of the hottest topics in the field today is the notion of teaching so-called twenty-first-century skills. These are the building blocks for innovation, and include such buzzwords as *creativity, collaboration, critical thinking, communication, empathy,* and *failure.* These differ greatly from the classic Western educational fundamentals of reading, writing, arithmetic, science, and so on, because they are so-called soft skills, more a series of behaviors than specific factual knowledge.

Today, these twenty-first-century skills are many of the same ones businesses say are in great demand, especially in the technology industry. A 2015 report by the National Bureau of Economic Research noted that jobs with increased social skill requirements have resulted in greater increases in employment and wages. This is ironic, because a big push of that same industry has been shifting educational funding away from liberal arts programs (writing, art, drama, social science) to science, technology, engineering, and math (STEM) courses, across all ages and school levels. Pursuing a degree in the liberal arts has been publicly derided as a dead-end path, even by President Obama (who holds a bachelor of arts degree from Columbia), but what these same businesses have come to realize is that a nation of skilled engineers and mathematicians is limited without creative, problem-solving minds to work with. Without creative skills, you are stifling original thought, which means that your job, and economy, can be outsourced and automated much more easily.

One of the most innovative companies with an analog solution to education is Twenty One Toys, which is based around the corner from the board game café Snakes & Lattes in Toronto. A small startup, Twenty One Toys was created by a young designer named Ilana Ben-Ari, whose business partner is a former high school teacher named

Ryan Burwell. When Ben-Ari was in university, her final project was to create a navigational aid for blind students at a school for the visually impaired. "I could have designed a BlackBerry with oversized buttons," she said. "But what if it got lost or the battery died?"

As a design student, Ben-Ari used a process called design thinking to approach problems, which was first developed by academics at Stanford University and the San Francisco design firm IDEO. Although designers looking to create solutions for their clients initially used design thinking, the methodology's popularity among other businesses and educational institutions has quickly spread. The first step of the design thinking process is empathy, essentially putting yourself in the shoes of the end user, who in Ben-Ari's case was a blind nine-year-old girl entering a new school for the first time. Defined as the ability to understand and share the feelings of another, empathy has become a hot topic in education and workplace research in recent years, especially when linked to technology. Studies have shown a marked decline in empathy in today's youth (one at the University of Michigan chronicled a 40% decrease in just 10 years), and the desensitizing effect of digital technology has been cited as a major reason behind this. The consequences of a less empathic population are dire: more narcissistic, more selfish, less cooperative, and potentially prone to violence.

As part of her research, Ben-Ari underwent orientation training at an institute for the visually impaired, and she realized that language was the key tool the blind used to learn about a space. So, she devised a toy that could help her blind student develop a common language of navigation with sighted peers. The toy was a wooden puzzle with five pieces: two shaped like honeycombs, and three shaped like arrows. Each piece had different notches and indentations on it, allowing them to interlock in near-limitless combinations, as well as textural markers, such as dots and rough ridges. Each box contained two identical sets of the pieces. The objective was to assemble one set, then vocally direct another user to identically assemble theirs. The set included two blindfolds, so sighted users could play with the same senses as blind users.

Ben-Ari found that her toy was equally popular with people who were not visually impaired, because it effectively taught them the intangible concept of empathy. This had much bigger educational applications than assistance for the blind, so Ben-Ari began marketing her newly dubbed Empathy Toy under the company name Twenty One Toys. "Essentially it's about a more human-centric approach to education," she said. Key twenty-first-century skills were difficult to practice or devise a formal curriculum around, but toys have been successful with teaching these abstract concepts ever since Friedrich Fröbel invented kindergarten in the nineteenth century. Fröbel, the father of early childhood education, included twenty "gifts," or learning toys, as part of his approach to play-based education. The name of Ben-Ari's company was a twist on this, creating the next generation of toys to continue Fröbel's legacy, with a focus on building twenty-first-century skills.

Ben-Ari and Burwell were adamant that the Empathy Toy was successful specifically because it was analog. "The type of complexities that empathy requires is often streamlined by technology," Burwell said. "So much of the complexities we once asked students to do, we've now designed tech solutions for." These solutions, like a Google search or a calculator, made arriving at answers easier, but in that simplified process, the deeper learning that came from solving a difficult problem—what Burwell saw as the core of education—was sacrificed for quantifiable, standardized answers. "With this," he said, holding up the Empathy Toy, "the goal is designed discomfort. We have this notion that easy-to-use devices are ideal, but in education, complexity is ideal." The toy forced users to tackle an evolving, messy, amorphous problem, and to do so cooperatively, using empathy to understand the needs of their fellow player and then create a common language that allowed them to complete the challenge.

In its first six months, Twenty One Toys had sold a thousand Empathy Toys to eight hundred schools in thirty-five different countries, as well as to businesses and other institutions. From elementary public schools to universities, NGOs, and management consultants, the beauty of the Empathy Toy was its endlessly adaptable nature, which taught a skill (really, a state of mind) that was universally beneficial, regardless of age, income, or area of focus. The Empathy Toy included teaching guides, sample games, and case studies to direct educators in potential uses, but in the end everyone adapted the toy to their own needs. A second grade teacher used it to teach students STEM concepts; a guidance counselor had deployed it to discuss emotions with troubled students; a business school professor taught leadership with it. Computer coding, website-user-experience design, language, literacy, ESL, occupational therapy, global health crisis management . . . the list of educational applications for the Empathy Toy was constantly growing. One fourth grade teacher used it to teach students what it must have been like for European settlers and Native Americans to try to communicate initially without a common language.

One day I joined Ben-Ari and Burwell at the modern campus of Sheridan College, one of the larger community colleges around Toronto. They were facilitating an Empathy Toy training session with twenty different teachers and administrators from the college's faculty. "This is a play-based workshop, so let's not just let words come out of our mouths," Burwell said, addressing the room. "One word for today is *tangible*. We call it a toy, but really it is a discussion tool that turns how we collaborate into a tactile experience we can talk about later."

There were three steps to the Empathy Toy learning experience: First, there was creative communication, framing how we talk to one another and what our assumptions are. Second was play, which took a tactile experience and applied it to these assumptions. The third, and most important, was discussing what occurred and why. This took the experience of the relatively simple game, and allowed the lessons from it to be applied to whatever problem the teacher was working on, regardless of subject.

There were endless combinations of gameplay, but Burwell started off with the most basic, which involved two blindfolded players and three pieces. "Any volunteers?" he asked. No one raised a hand. "Well, it looks like we have a real Canadian standoff here." Everyone laughed, and two teachers stepped forward and donned their blindfolds.

"Michael, okay, take the piece that's long and narrow and point its tip toward your chest," said the professor giving instructions to her fellow player.

"Okay."

"Next is the cog . . . oh, shoot, it's falling apart on me . . . okay, there we go."

"Got it."

"Okay, now slide the short arrow and position it at two o'clock on the cog."

"My two o'clock or yours? Wait . . . I think I have it."

"Then place the bumpy-faced cog and fit it into the short arrow . . ."

"On top of the arrow?"

This went on for five minutes, with many detours, false leads, and backward assemblies. But when they finally got the same shape and lifted their blindfolds, there was a big sigh of relief from everyone. The teachers all went back to their tables after this and played three different variations of the game. Some had two blindfolded players assembling with a third, sighted guide; another required players to tease out directions with questions; while a third had a guide reading out instructions that other players sent by text message.

One teacher who has used the Empathy Toy with successful results is Heidi Siwak, who teaches sixth grade at Dundas Central Public School, located about forty minutes outside Toronto. Dundas Central could be described as your average middle-class North American public elementary school. The school building was a mix of old and new, in various states of functional disrepair, and its cafeteria had been converted to class space a few years back. Siwak had been teaching at the school for fifteen years, and had earned a reputation around the region as an innovative teacher who had become adept with technology. In the past, her students had done projects through Twitter, cocreated an app with a class in New Zealand, and maintained blogs on a number of subjects. "I saw an increase in commitment, engagement, and interest in the kids," she said, referring to the start of her digital teaching efforts. "It was amazing." All of this happened at a school with limited computer resources. When I visited, the three computers in her class were nearly a decade old, and only one worked reliably.

A few years back, Siwak began to look more critically at the way she was using technology in the classroom. "I began to realize there were gaps in the pedagogy. You can have a whole project without doing the reading that takes it to a different level." The big change came in 2012, when Siwak attended the same integrative thinking seminar at the University of Toronto's Rotman School that I visited at the start of this chapter. With its emphasis on group problem solving, ongoing discussions, and concrete work, Siwak saw that integrative thinking and design thinking (which she also studied at Rotman) were far more innovative methods for teaching her students analytical skills than was anything requiring a screen.

"So much of what we learn benefits from being seen and discussed." A big part of this was the classroom space. The walls of Siwak's class were completely plastered with sheets of paper covered with notes, drawings, concepts, mottos, and other elements of the integrative and design thinking models that her students had been learning about all semester.

The day I visited, Siwak had presented her students with a design challenge that they would spend the week solving. The problem was a complicated one and a well-known design thinking case study: pediatric MRI scans were more expensive for patients who had to be sedated and anesthetized. The students had to figure out how to reduce these costs. Siwak showed the students a YouTube video of an MRI machine and how it worked, and then asked them to create a causal model, which is a word map identifying the various factual elements at the root of the problem. The kids broke into groups, and gathered around large sheets of paper with markers and pens. Siwak walked around the room, asking them questions to tease out potential causes.

"What is causing so many patients to be anesthetized and sedated?" she said.

"Because they're really scared," said one girl.

"Why?"

"The noise is really loud," said a boy.

"They don't know what will happen," said another.

"What if they're claustrophobic?" asked a girl.

"They might come out and find out they have an illness," said someone else.

"I mean, just imagine yourself in this dark, cramped, scary machine," said one boy with great dramatic flair. Siwak noted that they were exhibiting terrific empathy in examining the problem . . . a concept they had all learned about with the Empathy Toy a few weeks before.

The students then began turning the causes of the MRI problem into the questions that would lead to their design solutions: How could they get kids to lie still for the duration of the scan without relying on sedation? How could they reduce the cost of drugs so that it was cheaper to sedate patients? How might they make the waiting room experience, or the MRI itself, less stressful for families and patients? Siwak's students would create all sorts of inventive solutions over the week, from privacy curtains in the machines to MRI helmets that children could decorate themselves. They built prototypes for these out of cardboard, paper, Styrofoam, and glue, and then analyzed and criticized each group's invention. The only digital element was the brief video of an MRI Siwak showed the class.

"We get things out of this that we'd never get out of a digital experience," Siwak told me, as the class filed out. "When they're on, their ability to collaborate, ask questions, and come up with answers is astounding. They literally build off each other's ideas." The end result, in terms of learning, would build into their writing and communication skills, Siwak felt. It would leave them with the insight and ability to articulate their thoughts and collaborate, even if they never remembered anything about an MRI machine.

While design thinking can certainly use digital tools in its implementation, the best practices around it are thoroughly analog, and employ such tools as Post-it notes, large sheets of paper, LEGO, Play-Doh, and other pliable creative materials. Greenwood College School, which is one of the most expensive schools in Toronto, had all the latest state-of-the-art educational technology. Its students wrote assignments on Google Docs and received instant feedback and assessment online from their teachers. Leslie McBeth, a teacher who led a green industries class at Greenwood built around design thinking, saw this as technology's benefit, because it freed up time she would have spent in class doing administrative work and allowed her classroom time to be devoted to hands-on projects.

McBeth felt design thinking was particularly well suited to educating students for the modern economy and world. "The nineteenth- and twentieth-century model of education was to prepare workers to get a job and follow instructions," said McBeth, who grew up in a small industrial town, where that classic model of drill, test, and evaluate fed right into the factory jobs her friends and family still worked at. "The world has changed," she said, noting that the days of guaranteeing a successful future by finding a job where you followed instructions were largely gone, thanks in part to technology. "You need to think for yourself today. That's how you find success. To me, that shift is the purpose of education: to be critical thinkers and creative problem solvers."

I visited McBeth's eleventh and twelfth grade classes, which were both working on prototypes for projects they had approached through design thinking. One was a revitalization scheme for Toronto's waterfront, and the other was creating an indoor agriculture system. The students were producing all sorts of creative solutions, from elaborate models of their waterfront developments to fish farms where the fish's own waste would fertilize the plants that cleaned the water. It was loud, messy work. At one point, three girls were hand-sawing a piece of lumber balanced between two desks, and sawdust quickly coated their preppy uniforms and hair.

With a few exceptions, all the students said they preferred to work without computers on this type of project. They felt they had more creative freedom, were less distracted, could be more accurate to their vision, and gained a better understanding of the scale and materials involved. It also seemed more fun. The groups building models and contraptions around the room were laughing and joking as they glued and taped and cut and broke things. The only ones working on computers were two girls who gave up on a model and decided to make an app instead. They sat side by side, quietly checking out the pricing options on various app-building websites, flipping over to Facebook whenever McBeth was out of eyesight.

One of the key benefits to teaching a method of analog problem solving, such as integrative thinking or design thinking, is that it forces students to challenge their own assumptions and ideas. Students weren't graded on the final results of their solution, but on the way they approached the problem, collaborated as a group, and adapted along the way. There were no right answers in these projects. I saw this in action one day at the University of Toronto's Rotman School of Management, where third-year undergraduate students in the business design practicum class were tasked with creating disruptive new models for the school's MBA program, which ranked highly in Canada but lagged in international standing. What could the Rotman MBA look like in 2024 if there were no constraints?

Stefanie Schram, who ran Rotman's design program, told the students to look beyond the data they had gathered in surveys and research, and to really focus on the needs of their customers: Rotman's future MBA students. "Here in business," she said, "we have a tendency to fall in love with data and believe data is the whole truth, especially today with Big Data. But you can't only describe the world with quantitative data. Data only show the past."

Like Siwak and McBeth's elementary and high school classes, the university students were creating prototypes for their ideas, from physical models to illustrated comic strips, which they would then go upstairs and present to current Rotman MBA students and receive feedback. Schram had encouraged wild and imaginative ideas, and these included an MBA that consisted only of internships and online classes, an international MBA Olympiad, round-the-clock online faculty-on-demand, and the Flaming Sword, a choose-your-own-adventure video game that automatically selected a student's courses.

All the proposals dispensed with the traditional framework of Rotman as a physical school with classrooms, professors, and classes five days a week. "We're not going to waste your time in lectures," said one of the students from the group Designed Sealed Delivered. "It will be like the Khan Academy," he added, referring to the popular online mathematics lecture series. After all, they were building the school of the future, and as Christopher Federico mentioned the last time I was in this very building, virtual schools seemed to be that future.

But when the groups presented their prototypes to current MBA students, that future didn't look so certain.

"How are you going to get people motivated if they can't come to class?" asked one woman, who had previously worked in banking.

"Don't you know camaraderie and the network you build is the whole point of doing an MBA?" asked a student from India.

"I hate e-learning," said one woman, "plus I don't come from a business background. I wouldn't meet three quarters of my class if this was online. I learn specifically by going to class and by asking questions of students and professors."

The undergraduate business students had run up against the hard reality of creating the school of the future. Just because something seems like the obvious way forward, that doesn't mean it is. They based their assumptions on the fact that remote, digital learning was *possible*, but they didn't assess how well that model actually worked, or whether it was actually desirable for the MBA program at Rotman. And if recent history is any indication, the answer is a firm no.

One of the greatest promises and failures of the educational technology movement has centered on the massive open online course, or MOOC. Although fundamentally not that different from printed correspondence courses, recorded audio- or videotape lectures, or other distance-learning initiatives of the past century, MOOC emerged as a new term in 2008, when advances in streaming video technology and cloud computing made it possible to deliver lectures online in real time. A number of universities, colleges, and other institutions had previously experimented with putting some courses, lectures, and even degrees online, but MOOC fever really hit in 2012.

That was the year the artificial intelligence researcher Sebastian Thrun (creator of Google's self-driving car) and his partner Peter Norvig (Google's director of research) posted their introductory lecture on artificial intelligence at Stanford University online. Surprisingly, their lecture attracted more than 100,000 views, and the two launched the MOOC company Udacity, which along with its competitor Coursera and a handful of others promised to usher in a revolution in the way the world learned. Thrun predicted that MOOCs would be so disruptive, in fifty years only ten institutions in the world would provide higher education, and his would be one of them. Apparently a lesson about Icarus's journey to the sun wasn't included on the Udacity syllabus.

Udacity signed up hundreds of thousands of students, and universities quickly scrambled to jump onboard the MOOC train, posting their own courses online and offering up virtual degrees. At the start of 2013, Udacity signed an agreement with San Jose State University to launch a series of fully accredited online courses for the school. If successful, the partners argued, the courses could pave the path for a wider rollout of online degrees across the California state university system. The deal had been driven at the behest of California's governor Jerry Brown, who said there was a crisis with students entering college without the adequate requirements.

The trial was a massive failure. Within six months the university paused its MOOC experiment, citing dismally low completion rates for those who enrolled, and mediocre achievement rates for the few who managed to finish. The results caused Thrun to denounce many of Udacity's courses as "lousy." None of this should have come as a surprise. Online courses have consistently failed to engage users. Even Thrun himself acknowledged this when signing the deal with San Jose State, noting at the time that 90 percent of those who signed up for online courses typically dropped out, something he'd hoped, and failed, to change. Studies of online courses, from advanced classes affiliated with brick-and-mortar colleges and universities to charter school experiments from elementary to high schools, have all given online education an F. Yes, you can watch a fascinating lecture at home in your pajamas at any hour, but you are more likely to drop out of your class, and perform worse, and learn less, than your peers who are sitting in a school listening to a teacher talking in front of a blackboard.

There is a reason why Harvard and the University of Toronto and my old high school are still full, and why no one I went to school with would trade their years there for a MOOC or online degree, any more than they would for a correspondence course. That reason is teachers.

Teachers are the key to analog education's past, present, and future, and no technology can or should replace them. Not because they have the most knowledge, but because without them, education is no more than facts passed back and forth. If you want facts, go read a book. If you want to learn, find a teacher.

"Teaching and learning is a relationship between teachers and students," said Larry Cuban, the Stanford professor of education who witnessed the MOOC debacle unfold on his campus. "Relationships are analog. For those who push on technology, they interpret teaching and learning not as a relationship, but as a delivery for information. Education is not seen in relational terms at all. It is seen as a way of getting more access to information and having communication in ways not available before, and that isn't relational. I was a high school teacher and then a professor and then a superintendent. It became obvious to me that the whole basis of learning between the young and adults is anchored in the relationship between those people. A teacher has a relationship with a group of students. It is those independent relationships that is the basis of learning. Period."

Analog education, which happened in classrooms between teachers and students, and between students and other students, was more than just the transfer of data. That was the basis. But what teachers did, and could only do in the flesh-and-blood, person-to-person environments we call schools, was to take that raw information and mold it into knowledge.

When I think back on the twenty years I spent in school, what sticks with me isn't any particular subject, learning tool, or classroom. It is the teachers who brought my education to life and drove my interest forward, so that my passion for learning continued, despite the long days, the hard chairs, the difficult problems. These women and men were giants. They were underpaid, and they put up with all sorts of crap, but they made me the person I am today vastly more than the facts they taught. That relationship is what digital education technology cannot ever replicate or replace, and why a great teacher will always provide a more innovative model for the future of education than the most sophisticated device, software, or platform.

"For seven years, I have been teaching grades five and six here," Benjamin Peebles said the morning I visited his classroom at the Jackman School, while his students filed in. During that time, Peebles had been heavily involved with research into new digital learning technologies the University of Toronto was testing. "One thing that I have taken away is this: whatever technology is being used, the success

9

The Revenge of Analog, in Digital

O ne *Tuesday afternoon, Scott Unterberg* walked into a small, windowless room in Adobe's large San Francisco office and turned the lights on. As Unterberg straightened the Tibetan prayer flags hanging from the ceiling and distributed round floor cushions and Adobe Meditation–branded blankets, thirty coworkers from various divisions of the company entered the room, though not before they left their shoes, phones, and laptops in a neat row outside. The walls were decorated with forest wallpaper and stock photographs of people sitting in the lotus position. In the center of the room, seven smooth stones were neatly stacked into a tower.

Unterberg is the project manager in charge of Adobe's Creative Cloud suite of programs, which include such titles as Photoshop and Illustrator, but each day at three he opens up this room to lead the company's Project Breathe meditation session. Although he has been working in software his whole career, Unterberg is also trained in traditional Buddhist meditation, and has traveled often to India and Tibet, refining his practice. In 2008, several of his Adobe colleagues in the Flash software division asked Unterberg if he could teach them to meditate. Unterberg and a few members of the Flash team commandeered an empty meeting room once a week, where they meditated for fifteen minutes.

"Word of mouth spread quickly," Unterberg said. The dozen initial members swelled to as many as seventy, and the group gathered wherever they could get space, including an old bank vault that was now a conference room. Project Breathe evolved from this initial group, and though it now has a more formal name and company support, the most tangible benefit is a permanent space for the daily meditation.

"Hi, everyone," Unterberg said softly, sitting down cross-legged after someone dimmed the lights. "Brian, can you time us?" Unterberg picked up a small brass bowl and struck it with a wooden mallet, making a gong sound. Brian started the timer on his phone, which was the only device allowed in the room. Then we just sat. Five minutes later, Brian's timer pierced the silence. Everyone stretched, rolled their necks, and made loud exhalations. "Okay," Unterberg said, sitting up straight and striking the gong after thirty seconds. Brian reset the timer. Silence and stillness. Ten minutes later, the timer beeped, and Unterberg struck the gong one last time. "Thanks, everyone. See you tomorrow."

Over the past few years, Unterberg's Project Breathe meditation sessions have slowly expanded from this one San Francisco office to all of Adobe's offices worldwide. Adobe is by no means exceptional in this respect. Meditation and its broader umbrella movement, mindfulness, have become practically mandatory at the leading companies in Silicon Valley. Google's Search Inside Yourself program features regular meditation classes, and the company even has a purpose-built labyrinth for walking meditations. Facebook and Twitter both have meditation rooms in their offices, something that is now even found at hedge funds and banks. Zen masters, monks, and mindfulness gurus are as in demand in Silicon Valley as personal trainers and java script coders, and Unterberg himself has consulted with Yahoo!, Microsoft, Salesforce, SAP, and others (all entirely unpaid, as part of his Buddhist teaching). None of this is terribly new. After all, Steve Jobs, the great tech guru of our times, regularly practiced meditation, and much of Silicon Valley's roots are tied into Northern California's counterculture, when yogis, ashrams, and organic food went hand in hand with software design.

It is easy to dismiss all of this mindfulness talk as a PR stunt by giant, publicly traded corporations looking to appear more human; a lifestyle trend that will probably pass like cold-pressed juicing, or worse, as pointless hippie bullshit driven by Bay Area baby boomers assuaging their guilt for selling out. But scratch the surface beyond the "ohmm" chants and black turtlenecks, and such programs as Project Breathe reveal a deeper truth about the digital technology industry and the people working there.

The digital world values analog more than anyone.

I discovered this early on while writing this book. While my interest as a journalist has focused on analog, I have also spent some time over the past few years investing in digital technology companies. These have ranged from small startups here in Canada to venture capital funds based in Silicon Valley. These investments have put me in touch with dozens of people in the digital technology industry, from sixty-year-old veterans of enterprise software companies to twenty-year-olds with an idea for another app the world doesn't need. Whenever I told anyone who works in digital technology about the Revenge of Analog, they immediately began speaking about their own deep fascination with analog. More often than not, they harbored a personal passion for analog things. By day they wrote code, but at night they collected vinyl records, were starting a craft brewery, played board games, or repaired old motorcycles.

What was more interesting was where these views on analog dovetailed with their work in digital. More and more I began to encounter individuals and even whole companies where analog tools and processes played a significant role in building digital software and hardware. In some cases, this came down to personal habits. Nearly every single startup founder, investor, and programmer I met with carried a well-worn paper journal that they used to take notes and make designs, despite having access to every available digital alternative. "This *is* my company!" one startup founder told me, cradling the black Moleskine notebook in his arms.

The more I looked into this, the deeper it went. I read articles about the lives of technology industry leaders who spoke about their personal aversion to digital gadgets with their families. Steve Jobs didn't let his kids play with the very iPads he created, Chris Anderson from *Wired* and *The Long Tail* set time limits on technology for his children, and Evan Williams, who cocreated the digital publishing platforms Twitter, Blogger, and Medium, lived in a technology-free house, with a huge library of books. Silicon Valley and San Francisco, the mecca of ed tech, were also home to the most analog alternative schools in the country, from screen-free Waldorf and Montessori schools to outdoor kindergartens and a wild warehouse I visited called Brightworks School, where the children of digital titans built their own classrooms with saws and drills.

As my research into *The Revenge of Analog* progressed, my fascination with analog's role in the very digital technology industry that had disrupted it only deepened. I had seen the advances online retailers gained by opening stores, and heard about the growing interest in print publications by digital publishers. The most revealing conversations I was having about analog were almost always with those who worked in digital technology, because they wrestled with these questions daily. If there was a future for analog in our increasingly digital life, perhaps those who crafted that world had a better sense of what analog's role might be in the postdigital economy. Perhaps Silicon Valley's use of analog, even employed in the service of creating more digital things, could reveal the potential of analog in all sorts of ways I hadn't even considered.

———◦———

S cott *Unterberg's Project Breathe grew* from an ad hoc group into a company-wide initiative thanks to an analog innovation called the Adobe Kickbox. It is a cardboard box with the title "Idea Construction Kit" printed below a picture of a fire alarm, and the words "Pull in Case of an Idea" written in bold. Inside is a pack of Post-it notes, cue cards that spell out step-by-step instructions for taking an idea from a whim to reality, a pack of coffee, chocolates, pens and pencils, a paper notebook, and a $1,000 prepaid credit card. "It is intentionally a very hands-on, tactile, nondigital thing," said Kush

Amerasinghe, a computer scientist and strategy executive at Adobe who helped create the Kickbox and is a regular at Project Breathe. "It's so you focus on the idea, and not get constrained by the nitty gritty of technology that'll lead you away from your thought processes. Programmers inherently have a bad habit of jumping into code and building when they get an idea," he said. "Once built, they become married to it, and it narrows their horizons."

Unterberg used the Kickbox to expand the potential of his meditations into something deeper at Adobe. "'Hey,' I thought, 'what if we change the way we build software?'" His theory was that the individual approach to software building had become somewhat constrained by the team-centric focus of a large corporation, such as Adobe, with layers upon layers of management and decision making. If people could unplug and meditate daily, even for fifteen minutes, perhaps that would lead to better ideas for Adobe products. There were meditation apps out there that promised this, but Unterberg felt that unplugging was the cornerstone of the whole project. It was the only way to create the necessary mental space to process your thoughts away from the constant cycle of digital communications. "It doesn't matter what your job is anymore, you're plugged in," Unterberg said. Since the company had moved most of its products to the cloud (versus selling boxes of software on disks), the cycle of product innovation and releases had accelerated from two years, to a few months, to every few weeks. Fifteen minutes of silence a day was a life raft amid a sea of deadlines, iterations, and expectations that washed over Adobe's workers in endless waves.

After the meditation ended, I asked those sitting around the room why they attended Project Breathe. Although the group ranged greatly in their job responsibilities, most cited the same reasons. Meditation gave them a mental reset at a time of the day when it was needed most. It allowed them to step away from the stream of ideas. To go from reacting to information to quietly processing it, or even putting it out of mind altogether. One woman called these fifteen minutes a "big brain dump," which felt to her like walking away from a big TV blaring at full volume. An art director named Cindy ultimately told me how Project Breathe led her to better ideas. "Anyone who has ever gotten an idea in a shower knows that the idea has to come to you," she said, "not because you focused on that one thing, but because you stopped there and just let it go."

It was real "Kumbaya" stuff, but Adobe is a software company, and in an industry built on quantifiable data, numbers talk. So, Unterberg successfully got funding to study the effects of Project Breathe on participating employees. Participants filled out several surveys, and nurses came each week to take the blood pressure and heart rate of those who meditated. Across the board, those who participated in Project Breathe saw improved health, decreased stress, and other quantifiable factors. Adobe is now rolling Project Breathe out to every office it has globally.

Mala Sharma, an Indian-born executive who is Unterberg's boss and has practiced meditation all her life, believes this simple program has effected an even deeper change in Adobe's corporate culture. "The norm for everyone here is that we live our lives in technology," Sharma said, as we chatted with Unterberg after the session. "Project Breathe disrupts technology. Not all the results are that tangible, because ultimately meditation is a personal growth activity. But the ability to digest and engage in tough conversations allows people to be more objective in their work." Sharma felt those who participated grew better at giving and receiving feedback, accepting differing viewpoints, and digesting data, rather than just consuming it.

Project Breathe was just one aspect of what Sharma called a more analog, human-centric approach at Adobe. Some company executives had banned PowerPoint presentations in meetings, capped the length of e-mail chains, and opened up office floor plans, all in an effort to increase real-time conversations between employees and management, who had become so accustomed to digital communications that speaking face-to-face felt almost sacrilegious. This had a direct impact on productivity for such things as product releases, because disparate teams were now encouraged to physically sit down and sort out a problem in an hour, rather than sending messages and documents back and forth over the course of days, and scheduling video-conferences weeks later to discuss them.

"That's a massive change," she said. "You break down the artificial barriers that we built, and relearn how to do what we've forgotten."

---

*M*editation may be a fairly esoteric application of analog thinking in Silicon Valley, but across digital technology companies, the investment in analog is most visible in the physical workspace. Tech company offices, especially startups, are often rightly ridiculed for their outlandish, almost kindergartenesque atmosphere. The cliché of Segway-riding, foosball-playing coder bros picking up free kale smoothies on their way to the free bike mechanic class may be overplayed, but the reality is actually even more surreal. During my week in San Francisco and Silicon Valley, many of the offices I visited were just a talking chair shy of *Pee-wee's Playhouse*. This might seem like a reflection of high tech's often juvenile, nerdy executives and founders (give a twenty-five-year-old geek hundreds of millions of dollars and an architect, and what do you expect?), but there is a deeper purpose at the core, and it comes back to analog's advantage in the workplace.

To explain this, I met with Primo Orpilla, cofounder of the architecture and design firm Studio O+A, at the recently renovated headquarters for Yelp, which occupies a dozen floors of a historic office tower in downtown San Francisco. Orpilla has been designing offices for digital technology companies for over thirty years, basically since the start of the PC era. His clients have included everyone from Cisco and Microsoft to Facebook, Uber, and PayPal. Although they differ greatly in size and location, O+A offices have a signature look and feel to them, which can roughly be described as a midcentury modern ski lodge saturated in color. They are unified by playful, dramatic light fixtures, bright graphic installations, geometric furniture, long stretches of natural wood, exposed raw architectural features, and plenty of windows. Some have fireplaces, others have indoor gardens, and all have large kitchens.

What you don't find at a lot of O+A-designed offices is an abundance of high-tech gadgets. I expected robot baristas, touch screens on every surface, and other peeks into the office of the future, but what I found at Yelp and the other companies I visited were workplaces that were intentionally analog. "We suggest not to bring too much technology in. It's omnipresent," Orpilla told me, as we walked through the reception area at Yelp, which was designed as a sort of general store evoking the company's merchant customers, complete with glass jars filled with Yelp pencils and a shiny bronze cash register. "When it's a techy-filled environment it's miserable." He characterized these spaces as cold. "We want it more tactile and rough," he said, "more authentic to the building and a material." In fact, Orpilla saw more offices jammed with the latest technology being built for nontechnology companies (banks, law firms, retailers) who want to show they are cutting edge than for actual high-tech companies.

All O+A offices are unified by several feature spaces that Orpilla and his team have designed specifically to encourage face-to-face conversations. These range from casual encounters between an engineer and a salesperson in line for the daily free buffet in the cafeteria (a Town Hall), to one-on-one problem-solving sessions in a small, cocoonlike dome (a Shelter), to a formal meeting of executives during a crisis (the War Room). "The whole point of all these spaces is to get you to put down your device and read inflections, read body language, and have meaningful conversations," Orpilla said, as we walked into what Yelp called its All Hands floor, which was a large cafeteria with polished concrete floors and exposed brick walls, racks of free snacks and cereal, and a wooden coffee bar at its core staffed by an in-house, nonrobotic barista. Around it were large communal tables and intimate booths, where different Yelp employees, most wearing company T-shirts and hooded sweatshirts, sat, talked, and worked. The obligatory foosball table was nearby, and Orpilla told me that they purposefully put a greater selection of better food down here compared to the break rooms on individual floors to force people from different departments together, and ideally, come away with new ideas.

In contrast to these open spaces designed for collisions, there were other spaces meant for quiet, often solitary contemplation. These included craft rooms, bike garages, and other spots where workers could tinker with tools and hands-on activities, called Workshops;

small rooms with two chairs, called Think Tanks; and quiet rooms called Libraries, which in some companies were filled with actual racks of books and magazines, and in others, like Yelp's, were just serene spaces decorated with bookshelf wallpaper (the sight of which never fails to depress me). Some companies had a room devoted to board games, and at Yelp, a huge coffee table in one area was occupied by a giant puzzle of the Death Star.

We continued our tour through Yelp's engineering floor with John Lieu, the company's director of facilities and real estate. Yelp's actual work areas were recognizable to anyone who is familiar with an open-plan office: rows of desks where employees worked at computers. What struck me was the sheer quantity of whiteboards in these spaces. There were whiteboards on walls, on wheels, on the backs of laptop screens, and even painted on some furniture. Surely one of the leading technology companies in the world could afford the latest digital smart boards or other interactive collaboration technology. That's what Lieu had anticipated when he designed the office, which indeed was built with large digital displays in place of whiteboards.

"I nearly had a revolt from the engineers," Lieu said, when this technology was unveiled. Whole departments threatened to quit if their whiteboards weren't reinstalled, but when Lieu reluctantly brought them back, he noticed an immediate impact on the way the engineers worked. Writing on the whiteboard brought engineers out from behind their screens, and enticed them to take risks and share ideas with others. "If it's all computer based, are you really collaborating?" he asked. "I mean collaborating emotionally and physically?"

Orpilla felt that as digital technology became more pervasive, all office design would increasingly move in the direction of intentionally analog spaces and features, to encourage and even force more interpersonal collaboration. These would increasingly be seen as necessities that fed the very success of workers and organizations. Silicon Valley companies were ahead of the game on this, because they were the most tied to digital technology and saw the benefit of an analog workplace more than others.

"I don't think it's decadence," said David Pescovitz, the head of creative at the digital publishing platform Medium, whose offices are in a flatiron-shaped building above San Francisco's Union Square. Medium's office features nap pods, a high-end espresso machine, weekly wine-and-cheese parties with famous guest speakers, and the daily free meals (which Pescovitz and I were enjoying for lunch, along with Gabe Kleinman, Medium's head of product marketing). Prior to joining Medium, Pescovitz was a partner at Boing Boing, a large technology blog that was run virtually, with everyone contributing remotely. Boing Boing's team only met in person once a year. "Reality is still where the action is," Pescovitz said between bites of Moroccan chicken stew and quinoa salad. "I think people in this industry, because they are so engineering focused, are constantly looking for tech solutions to problems that analog solutions can do better."

Kleinman told me about an application someone at Medium had created to send coworkers a virtual high-five animation when they did a good job. It was cute, but in a way he saw it as a disservice to those it meant to encourage. "The way you support someone in a work environment is just not the same digitally as actually slow clapping or giving someone a high five in front of a big group of people," he said.

An intentionally analog workplace mattered more to digital technology companies for two key reasons. The first, which I had seen at Yelp and Medium, was creating a strong, interpersonal corporate culture, bound by real relationships, in an industry where the nature of the work, and the tools used to do it, naturally lean toward isolation. Offices that appeared at first glance like adult daycares were in fact carefully designed to maximize analog interactions, with an eye on fostering the company's culture of innovation and ultimate productivity. "People think these perks and benefits are there to just attract people," said Everett Katigbak, the brand design manager at Pinterest. "Really, these should support the work people do."

Pinterest's large, lofty office, in the increasingly tech-heavy SoMa neighborhood of San Francisco, was certainly flush with cool stuff. There were hanging gardens, DIY coffee tables, Ping-Pong, foosball, giant roadside signs, shelves with old cameras and Nintendo systems, a massive communal kitchen, and a huge central wall, easily four stories high, covered in photographs of Pinterest projects. In a way, Pinterest's headquarters was sort of like stepping into a living version of its online community, which is the other main reason why overtly analog workplaces dominated digital tech companies. Unlike consumer hardware manufacturers, such as Apple and GoPro, whose digital products you can hold in your hand, software companies are by their very nature ephemeral. They may be powerful international brands, but they have little to no presence in the physical world, with the sole exception of their offices. These corporate headquarters serve as embassies to the analog world, where the virtual brand can transcend into the physical.

Katigbak's background was in print design, and he had taught letterpress printing to others at Pinterest. Letterpress is a printing process of arranging physical block letters and woodcut designs, applying ink to them, and rolling paper over the top of them to make an inked impression in the paper. It is thoroughly archaic and analog, but it produces images that have a decidedly handmade look. It also seems to attract many who work in digital technology, largely because the process is so hands-on. Chris Chen, a senior software engineer at Twitter, believes that his hobby printing books with nineteenth-century typography machines feeds directly back to the way he writes software, producing code that is easier to read, understand, and edit than that of other engineers.

In late 2010, when Katigbak worked at Facebook, he and another designer on the marketing team named Ben Barry set up some of their printing equipment in the corner of a Facebook warehouse. Jokingly, they called it the Analog Research Laboratory. At a company where the culture was notoriously tech-centric, it was initially just a personal outlet for hands-on expression. "Others at the company said 'We are a digital company and we communicate via digital means,' Ben and I just got a bit antsy and wanted to make stuff," Katigbak said. "Part of this was our frustration over an obsession with data and metrics. Early on it was an attempt to humanize the brand for an internal audience, and to humanize the user."

They began making signs for the workplace with slogans about Facebook's hacker-derived work culture: "If It Works, It's Obsolete," "Is This a Technology Company?" "Move Fast and Break Shit," and every possible variation on the word *hack* and its use in a phrase.

Employees began noticing these signs hanging on cubicle walls and in hallways, and requested their own. Eventually, word got around to Mark Zuckerberg, who asked that the two produce hand-printed signs for Facebook's annual app developer conference. When that proved immensely popular, the Analog Research Laboratory was brought into Facebook's corporate structure, with its own dedicated space (next to a woodshop), budget, and full-time staff.

The Analog Research Laboratory is situated right near the main visitors' entrance at Facebook's sprawling, million-square-foot Palo Alto campus. The whole place, which is a closed village of different buildings and connected outdoor plazas, has a bright yet tightly controlled, *Truman Show* vibe to it, and the posters and signs the Analog Research Laboratory produces certainly reinforce this. You can't walk five feet at Facebook without bumping into a sign extolling the virtues of hacking or the sense of community employees are supposed to share. Many outside Facebook have called the Analog Research Laboratory the company's propaganda factory, and it certainly feels as if a poster of Mark Zuckerberg swimming across the Yangtze River, or Sheryl Sandberg smashing a Twitter bird under her outstretched fist, could emerge from its printing press at any moment. But at a company so large, charged with managing a social network whose very definition is amorphous, sometimes a little propaganda is needed to keep the cadres motivated and on track.

"How do you drive a large community and continue fostering a culture of autonomy?" asked Tim Belonax, the Analog Research Laboratory's current principal designer, as we talked over kale salads at one of the company's huge, free cafeterias. "The mission of the lab is to provoke and instill creativity in people."

Today, the lab regularly hosts different teams from the company, who often come in to make their own motivational signs ahead of a big project. It is both a team-building exercise and a stress relief, but crucially a way of distilling the collective work of a team into a tangible, visible slogan. An executive with the company's design team told me that employees needed this grounding element, because it gave them a different, more permanent sense of accomplishment than anything they made online. The poster on the wall would endure long after the project it was inspired by vanished from the website.

The great advantage of social networks, such as Facebook, Twitter, and Pinterest is the ease of joining them. But as anyone with an inactive MySpace, Second Life, or Friendster account will tell you, the ease of digital adoption is a double-edged sword. Consumer loyalty to these services is shallow, at best. Online, the main way to keep users engaged is to dominate the space: be the biggest social network, have the best features, and become so enmeshed in people's lives that leaving will be a pain. This is why such companies as Facebook purchase upcoming social networks, such as Instagram and WhatsApp: to head off a potential suitor before they steal the throne. But the scattered shells of thousands of failed online communities—some fledgling startups and others once-global leaders, such as AOL—show just how difficult a digital community is to keep together.

Analog provides a potential solution to this. If social networks and online communities are able to transcend their virtual existence into some form of real-life interaction, they can build a genuine sense of belonging among users, and with that, the type of customer loyalty that protects against competitors over the long term. An excellent example is Yelp, which launched in 2005 as a community of consumers and merchants, primarily focused on restaurant reviews. A year later, Yelp's CEO Jeremy Stoppelman and Nish Nadaraja (then Yelp's brand manager) were looking for a way to motivate their nascent community to review more frequently and reliably. They decided to reward the most loyal and prolific Yelpers with elevated status, an Elite badge on their profile that would grant them access to regular events and parties in select cities where Yelp operated. These range from group meals at a new restaurant to elaborate festivals with live music, whacky themes, free food, drink, and swag galore. Yelp Elite Squad events foster a flesh-and-blood sense of community, as opposed to other social networks, where participation is entirely virtual.

Elites are selected for their dedication and enthusiasm. They are the Christian Coalition to Yelp's Republican party; a fervent base that maintains cohesion and drives activity on the site, which in turn allows Yelp to grow its network globally and attract more advertising dollars. If ordinary Yelp users are infantry soldiers in the world of restaurant reviews, then the Yelp Elite Squad are Marines: the few, the proud, the most liberal in their use of exclamation marks!!!!!!! "The nucleus of Yelp is that community," said Nadaraja, who now advises startups. "Anything coming in (advertising, sponsors, etc.) is all based on that." Such loyalty has given Yelp a competitive advantage over other review sites, such as Citysearch, Urbanspoon, and Zagat. Elite enthusiasm and participation drive others to contribute to the site, raising Yelp's value for local businesses, who are its paying customer base.

"It's counterintuitive for an Internet business to bring people together offline in an organized way," said Greg Sterling, an Internet analyst and consultant in San Francisco. "Bringing these people together in the physical world was smart, because it played off the philosophy of Yelp ('Real Reviews. Real People') and strengthened the community."

Another online community that engages with its users by analog means is Behance, an Adobe-owned global platform for designers and other visual creative workers to showcase their work. "Analog's product can survive more," Behance's CEO and founder Scott Belsky said over breakfast near his home in New York. "Because you can profit off limited access. That's positive to the net: for choice, competition, and prices." Belsky has published a Behance book focused on business lessons and leadership, organized an annual Behance conference on creative thinking, and even produced a line of Behance notebooks and stationery (called Action Method) designed to transform ideas into action. Belsky said that people who have purchased any of those physical products have gone on to become the most fervent ambassadors for the Behance community in ways that digital products (eBooks, online seminars, virtual notebooks) would never foster.

An example of this is the annual in-person portfolio review week Behance runs for designers in hundreds of cities all around the world.

The best portfolios in local chapters receive recognition, but more crucially, select participants get a literal token of appreciation from Behance in the form of a metal coin with the company's logo. The token is a simple, inexpensive, silly thing, but Belsky firmly believes it makes a big difference in strengthening Behance's community. "These coins have become an iconic thing," he said. "There's a scarcity factor of physical goods, while digital goods are totally worthless. When we decided to reward the top people, we knew it had to be something physical."

Belsky felt the ultimate benefit analog instilled in a digital company's corporate culture was friction. In digital technology, friction typically represents obstacles, archaic practices, and barriers to be overcome. "Digital workplaces are designed to be frictionless," Belsky said, effortlessly holding up his phone and using an Adobe app his team helped create to capture the colors in a bowl of fruit with a single tap. "But the analog world is all about friction," he continued, running his hand over the rough wooden table. "It is a friction-filled experience. Should we have a completely frictionless life? Creativity happens in clashes. In truth, it is friction that sparks creativity. Without friction, things simply go as planned."

One of the great promises of the information age was that advancements in communication technology would result in increased productivity. Studies have shown that has not occurred, but most people don't need academic data to realize this. They simply need to look at the e-mails piling up in their inbox, at the texts pinging away on their phone, at the office-wide Slack thread that is spiraling out of their control to understand that any technology built with the promise of productivity has the real potential to deliver an inverse result.

What some technology companies have done in response to this is limit technology itself. At Percolate, a New York company that builds software to manage marketing departments, this involved the adoption of firm rules around meetings, which had become long-winded, often grueling affairs. One of these rules banned all digital devices from company meetings. Noah Brier, Percolate's cofounder and CEO, said the rule arose because he consistently sat in meetings where one person spoke and everyone else pretended to listen while they responded to e-mails or texted. Not only was this rude, but the distraction increased the length of meetings drastically. Once Perco-late banned devices, the results were instantaneous. "It just makes it so people are actually paying attention. Meetings are shorter and more useful."

Other technology companies have adopted different techniques to achieve the same results. Orpilla had heard about a semiconduc-tor firm in Silicon Valley that actually created a meeting room with signal-blocking technology. Amazon, on the other hand, opted for a more analog solution. When Jeff Bezos gathers his executive team in Seattle, the entire meeting is structured around a six-page narrative memo that executives are responsible to write. Every person who enters the meeting spends the first half-hour quietly reading, and the discussion begins only once everyone has finished the memo. In an interview Bezos likened the experience to study hall, but he believed that making executives compose their ideas into a narrative format forced them to articulate those ideas more clearly than they would with PowerPoint slides.

Beyond meetings, technology companies have adopted policies to encourage more face-to-face communication. Scott Heiferman, CEO of the event organization service Meetup, has built his company's corporate structure around the value of the same face-to-face inter-actions that Meetup facilitates. On any given day there are thousands of encounters happening all over the world thanks to Meetup, from bird-watching groups in Bangkok to record swaps in Buenos Aires. But Meetup only has one office, which is in New York City, and all its employees work under one roof. When I met Heiferman there, he told me this was a conscious choice. "I don't do phone calls, I don't let people Skype into a meeting, and I'm going to eat lunch with my friends," he said, as we sat on folding Coleman chairs in a small meet-ing room made to look like a campsite, complete with walls covered in photos of Meetup hiking groups. Heiferman refuses to have im-portant conversations by phone, e-mail, or other digital means unless absolutely necessary. "We believe being a community means you're face-to-face a lot. I think friendships matter and ultimately so much is better when face-to-face happens," he said.

These companies are not turning to analog out of some *Mad Men*–inspired nostalgia for the way business was once done, or because the people working there are afraid of change. They are the most advanced, progressive corporations in the world. They are not embracing analog because it is cool. They do it because analog proves the most efficient, productive way to conduct business. They embrace analog to give them a competitive advantage.

———◆———

One day in San Francisco, I met John Skidgel, a user experience (UX) designer for Google, at the company's cafeteria overlooking the Bay Bridge. Over roast chicken and yet another side of braised kale (nothing, apparently, is more analog than a high-fiber diet), Skidgel, who previously worked at Adobe and YouTube, told me about his own analog initiative. As a designer, Skidgel had always sketched his first drafts on paper, but most of the other designers he encountered at Google were designing right into illustration software. In 2009, Skidgel developed an internal class at Google for other UX designers on sketching. Everyone who showed up was handed a paper sketchbook and three pens of varying thickness.

The course, which lasted several hours, began with the most basic sketching task: drawing a straight line. Skidgel opened my notebook and showed me how he taught this. "You want to draw toward you in one swift, vertical motion," he said, "making sure to use your whole arm, but not to lock your joints." I tried it, and yeah, it was a pretty straight line. Next came horizontal lines, dotted lines, shading, text boxes (write the text, *then* draw the box), buttons, and so on, until designers were able to sketch out pretty much all the functional aspects of Google products on paper. The goal was to enable Google's designers to focus on quickly and effectively communicating new ideas, without getting mired in the infinitely adjustable variables that design software allows. It was so effective that Skidgel's course is now taught to all Google UX and UI (user interface) designers worldwide, and hand-drawn, ink-on-paper sketches are now the company's standard first step in the design process.

"Drawing is really quick," Skidgel told me. "It's cheap . . . just paper and pen, and you don't get bogged down in the details. Sketches suggest something, they're not dictating. You don't worry about shades and font." Computer design software immediately looks real, and because of this, designers too often get caught up in precise but utterly pointless details. With hand-drawn sketches, even though they appear rough, the focus is on the idea and it can be adapted in seconds. When Skidgel inevitably encounters a designer in a class who expresses skepticism ("Isn't this what computers are for?"), he asks them what would happen if they found themselves in the elevator with Google's founder Larry Page. Here was a once-in-a-lifetime chance to finally tell the big boss about your revolutionary new idea, and you had twenty seconds to sell him on it. "You'd take out a pen and sketch the idea on a napkin and give it to him," Skidgel said. "You're not going to be able to do that on a laptop."

Designers I spoke with who worked at companies such as Twitter, Dropbox, and Pinterest gushed about the unrivalled superiority of whiteboards, Post-it notes, and paper to take ideas from the mind into a tangible place. They weren't replacing design software with paper, or avoiding it. Once the paper designs allowed an idea to evolve into a more concrete state, the process invariably moved to the computer where the design could be refined and tested. But when it made that transition to digital, it was more thought out, and frankly better than a design that began on the computer.

One of the most interesting companies to feel this change has been Evernote, the same cloud-based paperless note organization service that produces notebooks with Moleskine. When I visited the company's headquarters in Redwood City, California, its vice president of marketing, Andrew Sinkov, told me that when the company began in 2007, paper was a source of friction to be overcome. "We saw ourselves as a company taking people paperless," he said as we talked with Jeff Zwerner, Evernote's vice president of design.

Zwerner told me that Evernote's founder (and former CEO), Phil Libin, made a conscious decision in 2013 to take the company in the opposite direction of its virtual roots and those of its competitors. That year, Evernote opened a marketplace for physical products, including its special Moleskine notebooks and Post-it notes, desk accessories (pencil holders, laptop stands), and even bags. Not only has the marketplace been a success in terms of sales—more than $1 million worth a month, according to Zwerner in early 2015—but these physical products have resulted in increased use of the company's virtual service. In the Evernote Moleskine notebook's first year on the market, customers who purchased it used Evernote's cloud-based note management service 10 percent more than they did before. "People get excited about physical products. They get emotionally attached to things," Sinkov said. "When have you done that with an app?"

These physical products have also become useful reference points for the company's digital aesthetic. Previously, Evernote's design department did almost all of its work on computers. The problem, Zwerner told me, was that no one really knew what anyone else was working on at a given time, because it was hidden in hard drives. Now, Evernote's designers print out new designs for both physical and digital products, which they pin to a wall that spans the length of the office. "The policy now is: get stuff up so it takes flight," Zwerner said. "It gives the product and software designers a new frame of reference, a physical manifestation of the brand. They get three hundred and sixty degrees of feedback that's not just limited to how it is on a screen." He walked over to a new, top-of-the-line Xerox color printer, patted it gently, and told me that this printer had become one of the company's most valuable pieces of technology.

At the end of the day, profit and performance are the drivers of companies in Silicon Valley, and more broadly, the greater global technology business. If analog can provide an advantage, companies will adopt it, and there are several areas where its use is quickly growing.

One area was so-called curated content, which basically means things that are selected by humans from a larger pool of information, such as recommendations for articles to read, books to purchase, or videos to watch. Many have tried, and largely failed, to accomplish this entirely with algorithms. Netflix's algorithm has repeatedly recommended that I watch the Pauly Shore/Stephen Baldwin vehicle *Bio-Dome*; it is hard to imagine a clearer example of math's failing to capture the complexity of human taste. During the summer of 2015,

Twitter, Instagram, and YouTube all independently announced new features for recommending content that use actual human beings to sift, select, and edit the best posts from their massive troves of incoming data.

Another is security. Cybersecurity, by all quantifiable measures, is an oxymoron. From the largest corporations to the most sensitive government networks, any computer that can be hacked will be hacked. "Digital technologies, commonly referred to as cyber systems, are a security paradox," wrote Richard Danzig, the former US secretary of the navy, in a 2015 paper for the Center for a New American Security, which looked at cyber vulnerability in critical military and government command-and-control systems. "Their communicative capabilities enable collaboration and networking, but in so doing they open doors to intrusion." Danzig proposed the integration of analog safeguards into critical systems. These include placing humans into decision-making roles, employing analog devices as a check on digital equipment (physical switches, for example), and providing analog backups if digital systems are attacked.

This falls into a growing field of thought around systems design and artificial intelligence called human-in-the-loop, which intentionally integrates humans into a digital process to steer the computer along using their analog judgment. There are established applications in such systems as nuclear power plants or airplanes, but it is now being applied to more benign consumer software. A big proponent of this is Tom Hadfield, whose online shopping app, Fetch, uses a mix of human shoppers and artificial intelligence to effectively make purchases based on simple requests by text, e-mail, or voice. Hadfield calls it "bionic assistance," or artificial intelligence (AI) augmented with human-powered intelligence. "With human judgment, if you said 'Hey, Tom, buy me a pair of Nikes in size twelve, in either blue or white, or if those are not available, then in red,' that's a very complicated task for a computer to interpret, but for a human, it's easy," he said. "We're using AI when it makes sense, and human intelligence when it makes sense. It's really a combo of the two that enables us to do what we mean to do." Facebook has employed a similar technique in its personal assistant software.

Finally there is the fundamental nature of digital computing it-self, which many say is reaching its limits. Computer engineers have a real fear that the inevitable progress dictated by Moore's Law could hit a wall in the next decade as processors grow beyond our ability to feed them adequate supplies of power. Each time a processor makes a calculation, switching on a 1 or a 0, that action requires electricity, and the energy efficiency of digital processors has been relatively stag-nant, compared to their gain in speed. The most commonly floated solution to this, which is still in the early stages of research, is so-called analog computing. This would rely not on the exact binary calculations of 1's and 0's flowing through silicon chips, but on more approximate calculations, which recognize patterns while using far less energy. It is highly complex, futuristic stuff, but then again, so are self-driving cars. Many say that analog is nothing less than the future of computers.

Silicon Valley is an idealistic place, far more so than other indus-tries, such as finance or manufacturing. Although its technical roots lie in the postwar military/industrial complex, its soul and heart are intimately tied to the counterculture movement of the late 1960s and early 1970s. When startup founders stand on stages at technology conferences and promise to change the world, their sentiment is gen-uine, and their belief in the transformational power of technology for good is downright religious.

As digital technology has evolved into a more virtual pursuit, however, many feel that it has become increasingly disconnected from its physical roots and, more consequentially, the real, analog world beyond the technology industry itself. Silicon Valley, once the symbol of countercultural upstarts, is the new Wall Street. The hackers have become the establishment.

One way to wrestle this back down to earth is to ground the technology industry more firmly in the analog. "People are so attracted to [screens], and pulled into [screens], but we realize there's an impov-erishment of senses," said Blaise Bertrand, the director of industrial design at the design firm IDEO, the same company that helped create the design thinking method. "As human beings we are multisensorial. We have so many ways to capture the richness of experiences, but people are focusing their attention more and more on the screen, and other aspects of senses (our touch, our smell) aren't innovating." Bertrand predicted that the discussion of analog's ultimate benefit to the technology industry would soon become the dominant topic in Silicon Valley. Those who would build the technologies that really could change the world were the ones who readily acknowledged the limits of digital and the benefits of analog.

"There's no rational debate around this. The world is analog, and digital is always a representation," said Dan Shapiro, whose company, Glowforge, has built a 3-D laser cutting machine able to cut such materials as leather, wood, and cardboard with tremendous precision. Glowforge was just the latest startup Shapiro had launched, but he had a strong appreciation for analog, and had even created a popular board game called Robot Turtles, which taught kids computer programming. He likened working in a purely digital environment, such as software, to playing a video game on Easy mode. Analog was vastly more challenging and consequential, but also more rewarding when it worked. Analog demanded your respect, and if you earned it, the possibilities were that much richer than in digital. "Digital is not reality," Shapiro said, "it's the most convenient way with machines we have that we can approximate reality. Moving from analog to digital is always a process of throwing things away. And what we can get away with. Analog is always the source, always the truth. Reality is analog. Digital is the best we can do with the tools of the day," he said. "It's funny how often people forget that."

———————

On the morning I was booked to fly home from San Francisco, I took a detour on the way to the airport to speak with Kevin Kelly, best known as one of the founding editors of *Wired* magazine. Kelly belongs to a core group of techno-idealists who see digital technology as a force for ultimate good, and he pioneered some of the earliest online communities and social networks. In 2010, Kelly published *What Technology Wants*, a book on how technology shapes us as human beings, which a young computer programmer named Mike

Murchison strongly suggested I read, specifically for Kelly's thoughts on how our use of technology evolved, and how that related to the Revenge of Analog.

"It seemed to me as if no technologies ever disappeared," Kelly wrote in the book, noting that even antique farm tools, fountain pens, and candles were manufactured and sold today, having achieved a "beautiful uselessness":

> Technologies have a social dimension beyond their mere mechanical performance. We adopt new technologies largely because of what they do for us, but also in part because of what they mean to us. Often we refuse to adopt technology for the same reason: because of how the avoidance reinforces or shapes our identity.
>
> Groups or individuals will reject all kinds of technologically advanced innovations simply because they can. Or because everyone else accepts them. Or because they clash with their self-conception. Or because they don't mind doing things with more effort. People will choose to abstain from or forsake particular global standards of technology as a form of idiosyncratic distinction.

Kelly met me in a large study attached to his home, much of which he built himself. Immediately upon entering I was struck by two things: a giant model of a robot that stood 15 feet tall, and floor-to-ceiling bookshelves that rose two stories, with hundreds more books piled up along the floor. "We have an attraction to analog things, because we live in analog bodies," Kelly said, as I asked him what, indeed, digital technology wanted, and where analog technology fit into that. "The frequency, scale, range, and smoothness of analog things are appealing to us, and that includes the process of natural limitations. We can look at a table of numbers and get some meaning to it, but it is easier to see or feel something."

Kelly firmly believed the distinctions between analog and digital would soon fall away, as the technology improved beyond our current expectations, and delivered the performance of digital with the comfort and familiarity of analog. It was simply a matter of progress.

He wrote with an ink pen in a Moleskine notebook because the Livescribe digital pen he had wasn't yet there in its technological capability, but when the Livescribe's pen reached that state, Kelly's ink pen would retire to a drawer, presumably.

The most interesting project Kelly has produced in recent years is a giant book, measuring nearly 3 feet square, called *Cool Tools: A Catalog of Possibilities*, which he gave me as a parting gift. One of the first jobs Kelly had was editing a publication called the *Whole Earth Catalog*, which featured reader reviews of products and essays that were geared to the first generation of proto-hippie-hackers, back in the late 1960s and early 1970s. Think geodesic domes, experimental solar panels, and screeds against corporate America. Later, the *Whole Earth Catalog* expanded to include computers and software, but the rise of review-based e-commerce sites, such as Amazon, rendered the *Whole Earth Catalog* somewhat pointless, and it ceased publication. Kelly kept it alive with a blog called *Cool Tools*, which published a single review of a different tool every day, in the spirit of the *Whole Earth Catalog*. Kelly continued updating *Cool Tools* online, but kept feeling there was this gap, the final 5 percent of the experience, that online simply couldn't achieve.

"Twenty years later, I'd find myself late at night looking through some of those old catalogs, and be amazed that even with much of that info out of date, it was incredibly transfixing," Kelly said. "Something was going on in this out-of-date, moribund information that could mesmerize me and tell me something for hours on end. What is it? I realized it was the format. It was the oversized pages, the jumbled layout, this wide-view navigation system of turning pages, large pages with unrelated things." He collected all the reviews on the website, created some new ones, and published them in a giant book. "The reason I put this back onto paper was to recapture that missing 5 percent that the web couldn't do. That's basically what it was."

That night, back home in Toronto after a long flight, I unpacked my bag, hefted Kelly's *Cool Tools* onto my lap, and opened it up. "Self-published or not, it's crazy to make a book on dead trees in 2013," Kelly wrote in the introduction. "There will be no Kindle or tablet version. This thing is heavy, costly to ship. But it is totally exciting as well. You can judge for yourself if it works." I turned the page, began reading reviews, and three hours later finally emerged from its trance.

Cool Tools was easily the most incredible analog creation I had encountered over the course of my research: an unwieldy, homespun artifact that amounted, essentially, to someone hitting Print on a website, featuring the most random assortment of product reviews you could ever imagine: the best industrial shelving, hand winch, micro-flashlight, umbrella, extreme pogo stick, fertility monitor, the best books on wilderness survival, mud construction, and gardening with urine. Everything you could possibly imagine, from the Nest smart home thermostat to the best translation of the Qur'an, was in its pages. It was the catalog to end all catalogs: less a source for products and more a timeless window into the world of consumer culture.

I left *Cool Tools* on my coffee table for a few months, and without fail, every single person who opened it immediately got sucked in. Part of this is the content and its randomness, but I credit most of its quirky appeal to the sheer analog nature of the damn thing. It is huge, impossible to ignore, easy to navigate, and all just right there, in your hands, on big pages that sound like thunder cracking when you flip them. I considered it exhibit A in analog's revenge, but when I chatted with Kelly about it by phone a few weeks later, he wasn't convinced. "Right now, *Cool Tools* had to be on paper," he conceded. "But in fifty years that may not be true. Paper was appropriate for this time right now, but it may not be appropriate in the future."

To be honest, I was somewhat disappointed by Kelly's response, though not entirely surprised. He was among the most vocal advocates for the all-improving nature of digital technology, and his belief in its enduring progress was unshaken. Kelly saw the Revenge of Analog as a minority counterculture, amounting to perhaps no more than 5 percent of consumers, if that. In terms of the general movement of culture, this was statistically insignificant. However, Kelly pointed out that the hippies were also a tiny subset of the population, and their influence on mainstream culture, music, politics, and even the spirit of Silicon Valley was clearly outsized.

Epilogue

The Revenge of Summer

*L*ast spring, just as *I* began writing the introduction to this book, my wife e-mailed me the new technology policy from Camp Walden, the summer camp I attended growing up. The web page began with a glossy video of camp life, with campers and staff speaking about the benefits of a summer free of technology. Below the video was a list of devices the camp banned (phones, laptops, tablets, anything with an Internet connection) and those it allowed (older MP3 players, digital cameras, e-readers that had no wireless connectivity). There was also a brief statement from Walden's director, Sol Birenbaum:

> We want campers to experience nature with all their senses, and engage directly with each other without the separation of a screen. At Walden, we encourage children to develop their sense of accomplishment and well-being through hands-on activities, whether they are sports or play or dance or music. And we want them to get those hands dirty! Which is why we need campers to leave most electronics at home. Help us preserve "endangered tech-free time" by supporting our policies.

The word *preserve* caught my eye. Camp Walden was inspired by the writing of Henry David Thoreau, who retreated to the relative wilderness of Walden Pond in Massachusetts to meditate on life away from the urban experience. Ted and Elaine Cole, the couple who founded Camp Walden and who ran it until Birenbaum took over, strongly believed in Thoreau's message. A portrait of Thoreau (who looked remarkably like Ted Cole) hung in the dining hall, and any camper or staff who came to Walden for ten summers was given a book of Thoreau's poetry during the first night of camp in a teary ceremony. The book was called *In Wilderness Is the Preservation of the World*, which was a line from one of Thoreau's poems on walking.

It seemed clear that the echo of Thoreau had been in Birenbaum's letter. His winking use of the word *endangered* said this almost overtly: that technology was encroaching on our natural state, the same way factories and roads had encroached on it in Thoreau's day. And then there was that word *preserve*—which I'd heard repeatedly throughout my research on this book. So the day after writing the book's final chapter, I packed a bag (which included the very sheets I slept on as a camper) and drove up to camp.

———— ✦ ————

Camp Walden *opened in 1970* on 750 acres of land, three and a half hours northeast of Toronto. The camp spreads across the south side of Red Pine Lake, a dark green body of water with an island in its center, the whole place entirely surrounded by thick woods and steep hills. There are five self-contained units of cabins for different age ranges (Colors, Comics, Zodiacs, Seekers, and Counselors in Training), and several dozen all-purpose fields and buildings housing such activities as tennis, water skiing, sailing, drama, ceramics, archery, and so on. Walden can accommodate upward of five hundred campers in a summer, making it one of the larger overnight camps in Canada, but otherwise it is a typical summer camp, right down to the mosquitoes.

From the night I first attended a slide show about Walden with my parents as a nine-year-old (where Ted Cole sent me home with a vinyl LP of a camp sing-along, which I still own), I spent ten summers there, from 1989 to 1998. I learned to water-ski, kayak, and carry a canoe; learned to hate tennis, give and receive wedgies, appreciate Bob Dylan's music, and light fires. I had my first kiss on the hill by archery, and got my first paid job as a counselor there. At the end of each summer, when Ted Cole lit a giant wooden W by the waterfront and sang Joni Mitchell's "Circle Game," I cried along with everyone else. The people I met at Walden remain close to this day, including Adam Caplan, whose friendship sparked this book.

Very little had changed since I left Walden half a lifetime ago. The buildings looked the same; the water had the same metallic taste; the crickets chirped the same staccato tune. Towels and clothes still hung from the front of each cabin's laundry line, and music blared on loudspeakers throughout camp, announcing the next activity. Boys sprinted from place to place, seemingly for no reason other than they could, and girls made up songs while braiding one another's hair. Campers still read *Archie* comics and made macramé bracelets. They even dressed in the same outfits: Teva sandals, baggy Roots sweatpants, college T-shirts. The conversations I overheard could have been plucked from any summer over the past half-century.

Sol Birenbaum (whom everyone just calls Sol) was sitting behind his computer in the camp office when I arrived. Birenbaum is in his early forties, with the kind of earnest positivity you expect in a camp director. He is a stark contrast to Ted Cole, who ran Walden like a benevolent monarch. When the Coles sold the camp to Birenbaum and a partner in 2003, Birenbaum immediately began thinking about digital technology's potential impact on Walden.

"I knew we'd have to consider questions that Ted and Elaine never would have," he told me. E-mails, blogs, and cell phones were already widely used technologies, and first-generation iPods were starting to replace the CD players that campers and staff brought up. The Coles had restricted the use of all portable music devices to cabins since the Walkman's invention, but Birenbaum saw that the challenges would only mount as the technology improved, and a line had to be drawn at the borders of camp. "I love technology ten months of the year," he told me, noting that he actually studied computer science, "but for two months of the year, we want to be technology free."

Over the ensuing summers, most parents quietly supported Walden's no-technology policy. A few vocally complained, noting that their kid needed TV or video games to fall asleep, but Birenbaum held fast, even as more technology worked its way into the fabric of kids' lives: smartphones, iPads, Facebook, Instagram, Snapchat, etc. Walden wasn't averse to using technology. The office ran on computers, and select staff used laptops to create various program materials—scripts for plays, cabin schedules, videos for evening activities—Walden even owned a drone to take videos—but campers had zero contact with computers throughout the summer.

Early on, Birenbaum realized that demanding a screen-free camp from campers required a trade-off with their digitally savvy parents, who demanded to know what was happening at camp daily. While my parents were content to receive a weekly letter from me with a few promising scraps of information, today's parents are accustomed to instant feedback. So, Walden began a blog, and then a Facebook page where a dedicated photographer posts a steady stream of photos, videos, and stories from camp. "If parents want to live vicariously through their children, then I say, 'Have at it!'" Birenbaum told me. "But I have my limits. No, I'm not going to check your kid's hair because you think it's dirty in the photo. No, I'm not going to speak to their counselor because you saw one photo of them not smiling. But, the reality is, the more pictures I send out, the less phone calls we get. I wish parents didn't require this, but they do."

One of Walden's biggest challenges with technology was e-mail. Traditionally, campers wrote and mailed letters home, and parents did the same. But increasingly parents demanded the ability to e-mail letters to their children, so Walden devised a compromise. Campers handwrite letters on a form that is scanned into Walden's computer as PDF files and e-mailed home three days later. Parents can e-mail Walden with notes for their child, and these are printed and delivered to campers three days after being received. The delay of three days on either end was designed to intentionally mimic the same lag that Canada Post experiences, which is crucial, according to Birenbaum, to preserve something he referred to as "the transfer of authority."

"Let's say a kid is getting bullied in a cabin by another camper," he said, using a recent example. "If she writes an e-mail home on her phone, her mother reacts immediately, advising action to her daughter, and contacting me to remedy the problem. The mother retains authority. But with a six-day delay from the time the daughter sends her letter to the mother's response, the camper has to deal with the problem of the bully. Eventually, the camper realizes that 'Hey, maybe this eighteen-year-old staff member taking care of me is someone who I should talk with,'" and you suddenly achieve that transfer of authority from parent to counselor that is crucial for Walden's social cohesion. Birenbaum believes the elevated anxiety he's observed in this generation of campers is directly related to the constant hovering of their parents, who use digital technology to keep tabs on their children around the clock. They cannot surrender their authority. Many of the phones that Birenbaum has seized from campers over the past few summers were sent on the insistence of parents, who wanted to remain in touch.

Walden's no-screen policy remains a key selling feature for most parents, especially since the introduction of smartphones and tablets, which have impacted family dynamics tremendously. "These people didn't experience a true deficit with their kids until recently," he said, noting that these devices have brought the once isolated computer into every possible moment of a family's life: at the dinner table, in the living room, in the car, on vacation. "Suddenly, everyone has their eyes on their phones, and the parents see this and say, 'What am I going to spend this money for, if they are going to sit in the cabin and Instagram each other?'"

By the summer of 2012, sporadic cellular phone coverage finally began reaching parts of Walden. Birenbaum knew that full service was only a matter of time. Various overnight camps dealt with phones in different ways. Some allowed partial use of cell phones and the Internet at rest periods, while others were completely open to campers and staff using their phones at all times. In 2011, a survey by the American Camp Association showed that fewer than 10 percent of camps let campers use their devices. By 2013, that number had nearly tripled, and continues to grow. Other camps have cracked down hard, including one near Toronto where staff used a handheld metal detector to flush out contraband phones. Campers have become more resourceful as well. Many bring up two phones: a primary phone they keep well hidden and an older "burner" phone that they allow to be confiscated.

Walden's no-screen policy is enforced through a strong degree of trust. Staff may bring phones and devices to use on days off, but at camp they must check these into a secure locker room. At the start of the summer, Birenbaum tells campers about the Amnesty Box in his office, where contraband items that are voluntarily surrendered will be returned to campers at the end of the summer.

Midway through the summer of 2015, Birenbaum was alerted of a potential cell phone infiltration, when a photograph on Walden's Facebook page was "liked" by the very camper in that picture. A few days later, Birenbaum held back all the Seeker campers (the oldest group, 12 to 14 years old) in the dining hall after breakfast. He spoke about the screen-free policy's reasoning, and told the campers that each cabin in their unit would be subjected to a search of all personal effects until the phones were either handed in or discovered. A few campers voluntarily handed in the phones right away. Several more were discovered once mattresses were flipped, combination locks cut, and every inch of their bunks searched. Birenbaum sent each phone home in the mail, COD, in an empty envelope without so much as a note to parents. The phone's presence on their front step was clear enough.

I asked Birenbaum what he was ultimately trying to preserve by keeping Walden technology free. Was it the land, the cabins, and the lake, and leaving those spaces undisturbed by the outside world? Or were his efforts to keep the digital barbarians at the gate driven by a desire to preserve something deeper, that universal truth that not only made Walden what it was, but drove the Revenge of Analog in all its various forms?

Birenbaum didn't hesitate to answer. "We look at the heart of what we do, and it is interpersonal relationships," he said. Any debate about technology's use came down to a simple binary question: will it impact interpersonal relationships or not? "This camp could be wiped out by a meteor tomorrow, and we could rebuild across the road and we'd still be Walden," he said. What mattered were the relationships and the uniquely analog recipe that enabled their formation.

First, you place lots of people together, and have them relate to one another with the guidance of caregivers, who encourage and enforce mutual respect. Next, you mix in a program that creates various stresses, frustrations, and challenges that campers need to confront. This ranges from the simplest task of getting to breakfast on time to ten-day canoe trips in the harsh Canadian wilderness where twelve-year-olds might be expected to carry a 60-pound canoe on their head for a mile or more in the pouring rain, as blackflies gnaw at their ankles.

These situations eventually lead to individual perseverance and self-respect . . . what most people call character. And that character is the glue that allows the relationships built at camp to last a lifetime, as my own friendships formed at Walden have. "You go a bit out of your comfort zone, endure a little hardship, people push you and help you to succeed, and you end up with friendships, confidence, and an inner fortitude that ends in a sense of belonging to a greater, interdependent community," Birenbaum said. "This is one of the most basic aspects of the human condition."

————◆————

Tech slang refers to the analog world outside the computer as IRL: in real life. This is a tacit acknowledgment, even from the acronym-loving hacker crowd, that digital is not reality. It never was and never will be. The frustrating, rough, rainy world outside of our screens is the place where our bodies and minds are at their best, and the place where they are built and grow and change.

We have been living with personal computers for over thirty years, the web for twenty, and smartphones for a decade. For each gain that digital technology delivers (speed, broad connectivity, vast processing power) it sacrifices something analog (quiet, personal connections, contemplative thought). We spend much of our waking time staring at screens, punching into keyboards, tapping, and swiping. Our days are ordered by the rhythms of digital sights and sounds: the ping of that first e-mail upon waking and the glow of the screen as we fall asleep in bed. We are more keenly aware of the few moments in our daily life when we are not online: asleep, in the shower, in the dwindling areas without a signal.

"[T]he digital world also brings dysphoria—a low-level but constant heartbreak that is one of its most controversial side effects," wrote Virginia Heffernan in the *New York Times Magazine* in 2011, in an article exploring the driving force behind a growing affinity for analog. "I also believed that we'd be over our nostalgic fixation on analog culture and its totems very quickly. Even the manual typists and vinyl collectors would find eBay soon, or YouTube or fantasy football, and they'd be off and running. And yet it's still here, the persistent sense of loss. The magic of the Internet—the recession of the material world in favor of a world of ideas—is not working for everyone."

The world today seems driven by disruption for disruption's sake . . . a chessboard flipped into the air, over and over again, so you're always scrambling to make your next move, without the time to contemplate the consequences. "Move Fast and Break Shit," the posters at Facebook say. Okay, great, everything's broken. Can we slow down for a second and maybe fix it?

For increasing numbers of people around the world, in nearly every place where digital life has acquired a real and lasting presence, analog is now a conscious choice, requiring a greater cost, both materially and in terms of our time and mental capacity, than the digital default. And yet people increasingly elect it.

Why?

One reason is pleasure. Analog gives us the joy of creating and possessing real, tangible things in realms where physical objects and experiences are fading. These pleasures range from the serendipity of getting a roll of film back from the developer, to the fun in playing a new board game with old friends, to the luxurious sound of unfolding the Sunday newspaper, and to the instant reward that comes from seeing your thoughts scratched onto a sheet of paper with the push of a pen. These are priceless experiences for those who enjoy them.

Another reason is profit. The Revenge of Analog represents a growing postdigital economy for goods and services that require investors, retailers, and entrepreneurs to make it happen. There is money to be made in analog, whether you open a small record store or a large watch factory. For all the press around the success of Silicon Valley, the vast majority of our economy is still overwhelmingly analog, and these businesses more broadly benefit our society than those of more concentrated digital capital. As the business world increasingly focuses on digital, companies and individuals who can use analog in new and novel ways will increasingly stand out and succeed. Human input will become more valuable, and analog tools and practices—from note taking on whiteboards to translating digital experiences into the real world (such as retail stores)—will separate the leading businesses from the rest of the pack. That is because analog is a tool of productivity. Often the best tool.

"We assume that anyone who rejects a new tool in favor of an older one is guilty of nostalgia, of making choices sentimentally rather than rationally," Nicholas Carr wrote in *The Glass Cage*. "But the real sentimental fallacy is the assumption that the new thing is always better suited to our purposes and intentions than the old thing. That's the view of a child, naive and pliable. What makes one tool superior to another has nothing to do with how new it is. What matters is how it enlarges us or diminishes us, how it shapes our experience of nature and culture and one another."

As more research comes out about the effect of omnipresent digital technology, we are also choosing analog for our health. Screen time has been proven to sap concentration, increase stress and anxiety, wreak havoc with sleep patterns, and disturb a whole host of brain functions. This is especially true in young children, but we can see the effects for ourselves in our own lives: the stress that comes with checking our devices every few minutes, the feeling of sluggishness we get after hours staring at a screen, the ever-present sense that we are missing out on something, that we are a step behind the world. Analog allows us to step back from this, losing ourselves for an hour or afternoon in a spinning record or a Sunday newspaper, and reassert our sense of who we are.

Ultimately, analog pursuits connect us to one another in a vastly deeper way than any digital technology can. They allow bonds to form in real time and in physical spaces, which transcend language and our ability to communicate with just words and symbols. People go to Snakes & Lattes to play Settlers of Catan, or take an MBA at the University of Toronto's nearby campus, not so much for games or

the degree, but for the indirect and vastly more beneficial social relationships that form there, in ways that are nearly impossible online.
All digital can bring us is a facsimile—constantly improving, but still
fundamentally a simulation—of real life and all its IRL richness.

The MIT professor Sherry Turkle, who has devoted her career
to studying and writing about the impact of digital technology on our
lives, once wrote that sociable technology always disappoints, because it promises what it cannot deliver. "It promises friendship but
can only deliver performance," she wrote in her seminal book *Alone
Together*. "Do we want to be in the business of manufacturing friends
that will never be friends?" A machine taken as a friend, Turkle
wrote, demeans the very notion of friendship itself.

Digital technology has the ability to fill in awkward moments
in innumerable and highly entertaining ways. There is a widely held
assumption that the naked human interaction unplugging leads to is
something we no longer desire. That is the counterargument I most
often encounter when I describe this book. That technology is a force
for good, that its presence enriches our lives and our relationships
by capturing them and amplifying them in all sorts of fantastic ways.
This assumption is especially true for the younger generation: those
who have grown up with digital technology their whole lives, as all
the campers attending Walden have. "Look how much time these
kids spend on computers, phones, and other devices," people say. "It is
what they know. It is how they communicate. It is what they love." To
deny them the digital technology at the core of their life, they say, is to
ignore the reality of a world that has fundamentally changed.

What I found over the course of writing this book, however, is
the exact opposite. The younger someone was, the more digitally exposed their generation was, the less I found them enamored by digital
technology, and the more they were wary of its effects. These were
the teenagers and twentysomethings out buying new turntables, film
cameras, and novels in paperback. They were the students who told
me how they would rather be constrained by the borders of a page
than the limits of word processors. These kids revered analog. They
craved it. And they were more articulate about its benefits than was
anyone else I spoke with.

During a barbecue lunch at Walden, I spoke with several groups of campers from different age groups about their thoughts on Walden's no-screen policy. The youngest, an eight-year-old girl named Reily from the Color unit, told me that she missed her iPad a lot. But her friends Alona (9) and Reese (8) both said they stopped missing their computers and devices within minutes of arriving at camp. "I mean, it would be bad to bring my iPad up here because I wouldn't do activities," Reily said, considering her friend's positions, "but good because I love my iPad!"

At a picnic table a few feet away, a group of teenaged Seeker girls and boys were all united in their support of a tech-free camp. "If Walden allowed technology, I'd leave," said Sami, a girl with long brown hair. "You're so connected right when you're here, you don't even need to be online. Anyone who you'd text is in your cabin anyway."

Noa, her friend, told me that their whole cabin convened a group chat online shortly before the summer, and pledged that none of them would bring phones to Walden. "We wouldn't be talking in the cabin and doing our nails," she said, imagining the alternative. "We'd be on our phones in the cabin, looking at Instagram. It wouldn't be camp." The spirit of this carried back to the city during their cabin reunions several times a year. Whenever these girls met up in Toronto, they enacted a no-phone policy, re-creating camp's analog dynamic.

Of course, once the summer ended, these campers would turn on their phones seconds after they stepped off the bus and hugged their parents. But they also appreciated that analog had a valuable place in a life dominated by digital interactions, and they made a conscious place for it, at camp and at home. This held true for nearly everyone I spoke with for this book, from record store owners to workers at high-tech companies. No one, including myself, advocated a return to the predigital lives we once knew. No one was flinging their phones into lakes, or exclusively living off the grid. An entirely analog existence was unattainable and unattractive, but so was an exclusively digital one. What was ideal, and what lay behind the Revenge of Analog, was striking a balance between the two.

I wanted to speak with some of the Seeker campers whose phones had been confiscated two weeks prior, and I was led to one of the furthest cabins in camp. Inside I found three of the campers—Kyle, Jake, and Michael—each of whom had brought a phone to camp for a different reason. Kyle simply found it hard to separate from such an indispensable tool, especially considering it was how he listened to music and took photos. Jake and Michael used theirs for communication. "It was just easier and faster than writing letters," Michael said, noting that he wanted to talk with his mother.

How had their perspective on camp changed since Birenbaum took their phones away?

"I think it's good to come here and get a break, if you like it or not," Jake said. "You need it. A big reason why we come here is to get away from everything. Not just our phones, but computers, TV, video games, and all the other distractions."

"Sol's doing this for a reason," Kyle said. "To preserve camp. Not to be a dick. No one needs to pay $7,500 to sit in a cabin on their phone." Kyle said Birenbaum was trying to teach them a lesson about the role of technology in their life. "I noticed how attached we are to it," Michael said, nodding.

I asked them if they regretted bringing their phones, and to a man, they all said they did. Did they miss the phones? No, they all said in unison.

"It was the best thing that could have happened to us," Kyle told me.

I walked back to the parking lot, turned on my car, and plugged in my own phone. There was no service. I switched on the stereo, which offered half a dozen digital listening options (MP3s, podcasts, CDs, satellite radio, streaming music), and opted for a classic rock station. Neil Young's "Helpless" was playing, of course. I took one last look at camp, opened up the windows, and inhaled. The air smelled of childhood. A childhood whose essence was preserved, at least for now.

I turned off the stereo and my phone, pulled onto the highway, and listened only to the wind as I picked up speed all the way home.

Acknowledgments

While I owe each and every one of these people a firm, analog hug for their assistance in making this book possible, I hope that having their names sealed in ink will at least provide sufficient evidence of my gratitude.

First, thank you to my agent Robert Guinsler, at Sterling Lord Literistic, for having the continual belief in me to make this project a reality, and putting me back into the hands of the good folks at Public-Affairs. It is a continual pleasure to get to work with such people as Peter Osnos, Clive Priddle, Lindsay Fradkoff, Jamie Leifer, Tony Forde, Matty Goldberg, Melissa Veronesi, and the rest of the team, who impress me at every step of the way. Congratulations especially to those members of the team who had *Revenge of Analog* babies during the process of this book's publication, which proved especially fertile.

At the top of this list is my exceedingly talented editor Benjamin Adams, whose patience, insight, and judgment are never in doubt, and who remains an absolute joy to work with. Since 2012, we have somehow produced two books (together) and four children (collectively), while retaining some semblance of sanity.

Countless people provided advice, connections, and a place to stay during the research and writing of this book, and I am truly grateful for all of it. David Katznelson and Jay Millar illuminated the world of vinyl for me, and Emily Spivak opened up the pages of the wonderful world of Moleskine, whose entire team was extremely welcoming in Milan. Grazie to Marco and Nicola for hosting me in Ferrania, and to Doc, Matthias, and Sally for their time in Vienna.

The entire crew at Snakes & Lattes are the best neighbors and board game gurus I could ask for.

Thank you once again to Ariadne, Aaron, Lucas, and Emily for a base of operations in London, and to Jeremy Leslie, Steven Watson, and others for sending me to the right ink-stained wretches of that lovely city. In New York's retail world, I am indebted to the fearless booksellers in that city, especially Chris Doeblin and Annie Hedrick at Book Culture. Thanks to the Shinola team, Kyle Polk, Amy Elliott Bragg, and Ben Blackwell, for everything in Detroit, and to everyone on the University of Toronto's various business and education faculties, for guiding me through the complicated world of education's future. Mike Murchison provided the inspiration for Chapter 9, Anne and Jeremy provided the lovely accommodations, and such people as Scott Belsky, Todd Krieger, and Rebecca Bortman got me into where they keep all that tasty free kale. Getting to return to the happiest place of your childhood is a blessing in itself, but being able to go back to Walden "for work" was a step beyond. Thank you, Sol, Jen, and the entire camp staff, for hosting me.

Much of the initial research found in this book was compiled by the talented Wendy Litner, who is equal parts brilliant and hilarious, and no doubt on her way to Hollywood right now, having just sold her script for the next great awkward-girl sitcom of our time. Wendy, once again, I am in your debt.

To David, Charles, Mark, Petar, Pamela, and all the other great folks at the Lavin Agency, thank you for taking this idea out into the world and helping me spread my wings beyond my usual audience of burly potato farmers.

Thank you to Jeremy Keehn at the *New Yorker*, and to the crew at *Bloomberg Businessweek*, for letting me explore these ideas in your pages.

Thanks to Reboot for kicking off my thinking about this, nearly a decade ago, atop a (luxury) mountain in Utah with some wonderful people, including unplugging guru Dan Rollman.

The Revenge of Analog was born out of my friendship with Adam Caplan, who remains not only one of my dearest friends, but also one of the people I can discuss anything with for hours on end. Adam, may your life continue to be as buoyant and whipped cream covered as a Herb Alpert album cover.

I owe so much of all this to Lauren, the love of my life, whose relationship with me has always been truly, deeply analog in the best possible sense. Thank you for your patience, your wisdom, and your encouraging me to write this book for the right reasons, and, of course, for being the best partner I could ask for, especially in raising our little IRL wonders.

Finally, I bow deeply in gratitude and awe for the guardians of analog who have kept its flame burning in their record stores, workshops, factories, studios, and minds during the darkest, most pessimistic times. This book is for you.

Selected Bibliography

INTRODUCTION

Embracing Analog: Why Physical Is Hot. JWT/Frank Rose, 2013.

Rushkoff, Douglas. *Present Shock: When Everything Happens Now.* Current, 2013.

Turkle, Sherry, and William J. Clancey. *Simulation and Its Discontents (Simplicity: Design, Technology, Business, Life).* MIT Press, 2009.

CHAPTER 1: THE REVENGE OF VINYL

Database of record stores found at recordshops.org.

Sales statistics courtesy of International Federation of the Phonographic Industry (IFPI) and the Recording Industry Association of America (RIAA), as well as Nielsen Soundscan, Record Store Day, and the Vinyl Factory.

Barnes, Tom. "Science Shows There's Only One Real Way to Listen to Music." *Music.Mic,* November 13, 2014.

Bartmanski, Dominik, and Ian Woodward. *Vinyl: The Analogue Record in the Digital Age.* Bloomsbury, 2015.

Bauerova, Ladka. "Czechs the Spin Kingpins in Global LP Revival." *Bloomberg,* February 11, 2015.

Blacc, Aloe. "Aloe Blacc: Streaming Services Need to Pay Songwriters Fairly." *Wired,* November 5, 2014.

Crane, Larry. "Jack White Is No Fan of Digital Audio." *Tape Op Magazine,* March 2011.

Graham, Jefferson. "Who's Making Money in Digital Music?" *USA Today,* February 15, 2015.

Greenwald, David. "Does Vinyl Really Sound Better? An Engineer Explains." *Oregonian,* November 19, 2014.

Grundberg, Sven. "A Penny for Your Song? Spotify Spills Details on Artist Payments." *Wall Street Journal*, December 3, 2015.

Guarino, Mark. "Pressing Plants Feel the Strain with Vinyl Records Back in the Groove." *Washington Post*, September 26, 2014.

Harding, Cortney. "Vinyl Gets Vital: A Classic Format Makes a Comeback." *Billboard*, November 17, 2007.

Harris, John. "Vinyl's Difficult Comeback." *The Guardian*, January 7, 2015.

Hasty, Katie. "Dave Grohl Talks Digital vs. Analog for Next Foo Fighters Album." *HitFix*, March 18, 2013.

Hochberg, William. "Just How Much of Musical History Has Been Lost to History?" *Atlantic*, September 26, 2013.

Hogan, Marc. "Did Vinyl Really Die in the '90s? Well, Sort of . . ." *SPIN*, May 16, 2014.

ICM Unlimited. "Music Buyers Prefer CDs, Vinyl and Cassettes over the Cloud." April 16, 2014.

Levy, Joe. "Jack White on Not Being a 'Sound-Bite Artist,' Living in the Wrong Era and Why Vinyl Records Are 'Hypnotic.'" *Billboard*, March 6, 2015.

Locker, Melissa. "A Fresh Sound: Whole Foods Starts Selling Records." *Time*, August 23, 2013.

"The Loudness Wars: Why Music Sounds Worse." NPR *All Things Considered*, December 31, 2009.

McDuling, John. "The Music Industry's Newfangled Growth Business: Vinyl Records." *Quartz*, July 11, 2014.

———. "The Vinyl Revival Is Not About Sound. It's About Identity." *Quartz*, January 9, 2015.

Oliphint, Joel. "Wax and Wane: The Tough Realities Behind Vinyl's Comeback." *Pitchfork*, July 28, 2014.

Paz, Elion. *Dust and Grooves: Adventures in Record Collecting*. Ten Speed Press, 2015.

Peoples, Glenn, and Russ Crupnick. "The True Story of How Vinyl Spun Its Way Back from Near-Extinction." *Billboard*, December 17, 2014.

Petrusich, Amanda. *Do Not Sell at Any Price: The Wild, Obsessive Hunt for the World's Rarest 78rpm Records*. Scribner, 2014.

Sottek, T. C. "Musician Jack White Praises Analog Living, Says 'There's No Romance in a Mouse Click.'" *Verge*, February 19, 2013.

"The Streaming Price Bible—Spotify, YouTube and What 1 Million Plays Means to You!" *Trichordist*, February 11, 2012.

Tingen, Paul. "Inside Track: Jack White." *Sound on Sound*, October 2014.

Van Buskirk, Eliot. "Vinyl May Be Final Nail in CD's Coffin." *Wired*, October 29, 2007.

The Vinyl Factory. "HMV Reclaims Top Spot as Britain's Biggest Physical Music Retailer." January 16, 2015.

———. "Turntable Resurgence: 240% Spike in Record Player Sales at John Lewis." May 5, 2015.

Whitwell, Tom. "Why Do All Records Sound the Same?" *Cuepoint—Medium*, January 9, 2015.

Chapter 2: The Revenge of Paper

Carbone, Ken. "Unify, Simplify, Amplify: How Moleskine Gets Branding Right." *Fast Co.Design*, March 28, 2011.

Chemin, Anne. "Handwriting vs. Typing: Is the Pen Still Mightier Than the Keyboard?" *The Guardian*, December 16, 2014.

Courtice, Craig. "The Cult of the Moleskine." *National Post*, November 11, 2006.

Francese, Alberto. "Moleskine: Brand and Model to Catch Target Market Growth." Banca IMI, March 24, 2015.

"Hacking a GTD Moleskine." *Lifehack*, January 2007.

Harkin, James. *Niche: The Missing Middle and Why Business Needs to Specialize to Survive.* Abacus, 2012.

Horowitz, Jason. "Does a Moleskine Notebook Tell the Truth?" *New York Times*, October 16, 2004.

Jabr, Ferris. "The Reading Brain in the Digital Age: Why Paper Still Beats Screens." *Scientific American*, November 1, 2013.

Levitin, Daniel. *The Organized Mind: Thinking Straight in the Age of Information Overload.* Dutton, 2014.

Martin, Claire. "Moleskine Notebooks Adapt to the Digital World." *New York Times*, April 18, 2015.

Mayyasi, Alex. "Is Moleskine Inc Replicable?" *Priceonomics*, March 22, 2013.

Mediobanca Securities. "Italian Wake-up Call." March 25, 2015.

Mueller, Pam, and Daniel Oppenheimer. "The Pen Is Mightier Than the Keyboard: Advantages of Longhand over Laptop Note Taking." Association for Psychological Science, 2014.

"On the Cards." *The Economist*, March 14, 2015.

Raphel, Adrienne. "The Virtual Moleskine." *New Yorker*, April 14, 2014.

Seward, Zachary. "Everything You Need to Know About Moleskine Ahead of Its IPO." *Quartz*, March 20, 2013.

Walker, Rob. "Look Smart." *New York Times Magazine*, June 26, 2005.

Weiner, Eric. "In a Digital Chapter, Paper Notebooks Are as Relevant as Ever." NPR, May 27, 2015.

Young, Molly. "A Pencil Shop, for Texting the Old-Fashioned Way." *New York Times*, May 19, 2015.

CHAPTER 3: THE REVENGE OF FILM

Film industry sales figures taken from Film Ferrania investors' presentation, compiled from Fujifilm, Agfa, Ilford, and selected articles and industry reports.

Fujifilm financial and sales information courtesy of annual/quarterly reports.

Japanese camera sales statistics courtesy of CIPA.

Ager, Steve. "Film Didn't Die with Kodak's Chapter 11." *Financial Times* video, January 4, 2015.

Bonanos, Christopher. *Instant: The Story of Polaroid.* Princeton Architectural Press, 2012.

Cade, D. L. "Teens 'Turning Their Backs on Digital' and Flocking to Polaroid, Says Impossible Project CEO." *PetaPixel*, November 9, 2014.

Hardy, Quentin. "At Kodak, Clinging to a Future Beyond Film." *New York Times*, March 20, 2015.

Kirn, Walter. "Remembrance of Things Lost." *New York Times Style Magazine*, April 12, 2015.

Klara, Robert. "How One Man Hopes to Restore the Legacy of Kodak." *Adweek*, October 20, 2014.

Lanier, Jaron. *Who Owns the Future?* Simon & Schuster, 2014.

"Leadership in Black and White—How a Manufacturer Profits in a Declined Analogue Film Industry." vivianeli.com, April 15, 2015.

Lomography. *LOMO Life: The Future Is Analogue.* Thames & Hudson, 2013.

"Minnesota's Pohlads Acquire Polaroid Majority Stake." *Pioneer Press*, December 27, 2014.

Phelps, David. "Five Years Later: Tom Petters' Ponzi Scheme." *Star Tribune*, September 23, 2013.

Renfroe, Don. "Fans of 'Analog' Photography Keep the Faith." *Des Moines Register*, January 19, 2015.

Rizov, Vadim. "Kodak's Back in Action and Making Film Stock Again." *Dissolve*, September 4, 2013.

Swart, Sharon, and Carolyn Giardina. "Film Fighters, All in One Frame." *Hollywood Reporter*, December 17, 2014.

Zhang, Michael. "30% of Film Shooters Are Younger Than 35, Says Ilford." *PetaPixel*, February 4, 2015.

CHAPTER 4: THE REVENGE OF BOARD GAMES

Hobby games market statistics and figures courtesy of ICV2.

Curry, Andrew. "Monopoly Killer: Perfect German Board Game Redefines Genre." *Wired*, March 23, 2009.

"Dispatching Obscene Boxes." *The Economist*, June 9, 2014.

Duffy, Owen. "Board Games' Golden Age: Sociable, Brilliant and Driven by the Internet." *The Guardian*, November 25, 2014.

Ewalt, David. "Fantasy Flight Games Merging with Asmodee." *Forbes*, November 17, 2014.

Ewalt, David M. *Of Dice and Men: The Story of Dungeons & Dragons and the People Who Play It*. Scribner, 2013.

Furino, Giaco. "Board Game Creators Are Making Assloads of Money on Kickstarter." *VICE*, September 17, 2014.

Gilsdorf, Ethan. "Board Games Are Back, and Boston's a Player." *Boston Globe*, November 26, 2014.

Kuchera, Ben. "No One Is Getting Rich from Exploding Kittens' $8.7 Million Kickstarter." *Polygon*, February 25, 2015.

Lagorio-Chafkin, Christine. "The Humans Behind Cards Against Humanity." *Inc.*, January 6, 2014.

Moulder, Stuart. "Boardgames: The Latest Analog Craze." *GeekWire*, November 27, 2014.

O'Neil, Lauren. "Cards Against Humanity Sells 30,000 Boxes of Actual Poop to Mock Holiday Consumerism." *CBC News*, December 15, 2014.

Ochs, Rhiannon. "Kickstarter Killed the Board Game Star." *Whose Turn Is It Anyway?* December 10, 2014.

Raphel, Adrienne. "The Man Who Built Catan." *New Yorker*, February 12, 2014.

Schank, Hana. "How Board Games Conquered Cafes." *Atlantic*, November 23, 2014.

Summers, Nick. "Cards Against Humanity, the Most Offensive—and Lucrative—Game on Earth." *Bloomberg Businessweek*, April 24, 2014.

Thai, Kim. "Board Games Are Back." *Fortune*, July 10, 2009.

Wingfield, Nick. "High-Tech Push Has Board Games Rolling Again." *New York Times*, May 5, 2014.

CHAPTER 5: THE REVENGE OF PRINT

Magazine statistics courtesy of Launch Monitor (Samir Husni).

Battan, Carri. "Is Vice Getting Nice?" *Daily Intelligencer*, April 1, 2015.

Biasotti, Tony. "The California Sunday Sets Out to Win the West." *Columbia Journalism Review*, October 21, 2014.

Bilton, Nick. "In a Mother's Library, Bound in Spirit and in Print." *New York Times*, May 13, 2015.

Bilton, Ricardo. "Why So Many Digital Publishers Are Flocking Back to Print." *DigiDay*, March 10, 2014.

Burrell, Ian. "Looks Good on Paper: Forget Tablet Editions—A New Wave of Young Independent Publishers Is Producing Wonderful Hard-Copy Titles." *The Independent*, February 19, 2014.

Carr, David. "Print Starts to Settle into Its Niches." *New York Times*, January 5, 2014.

Catalano, Frank. "Paper Is Back: Why 'Real' Books Are on the Rebound." *GeekWire*, January 18, 2015.

Changizi, Mark. "The Problem with the Web and E-Books Is That There's No Space for Them." *Psychology Today*, February 10, 2011.

Herships, Sally. "More Than 800 Magazines Launched in the Last Year." *Marketplace*, December 12, 2014.

Jackson, Jasper. "Guardian CEO: 'The Idea We Will Survive by Becoming a Technology Company Is Garbage.'" *Media Briefing*, December 9, 2014.

Milliot, Jim. "For Books, Print Is Back." *Publishers Weekly*, January 2, 2015.

Nowak, Peter. "Print Books Are Surviving—Even Thriving—in the e-Reader Age." *Canadian Business*, March 20, 2015.

Raphael, T. J. "Your Paper Brain and Your Kindle Brain Aren't the Same Thing." *PRI*, September 18, 2014.

Reese, Diana. "In Small Towns with Local Investment, Print Journalism Is Thriving." *Al Jazeera America*, April 29, 2014.

Sanders, Sam. "J.C. Penney Brings Back Its Print Catalog, After a 5-Year Hiatus." *NPR News*, January 20, 2015.

Silcoff, Mireille. "On Their Death Bed, Physical Books Have Finally Become Sexy." *New York Times Magazine*, April 25, 2014.

Tepler, Benjamin. "*Kinfolk* Magazine Takes Over the World." *Portland Monthly*, April 2, 2014.

UK Magnetic Influencer Survey 2015.

Van Meter, William. "A Fashion Magazine's Successful Business Model (Hint: It's Free!)." *New York Times*, March 4, 2015.

Wilkinson, Alec. "Read It and Reap." *New Yorker*, November 10, 2014.

Wolff, Michael. "How Television Won the Internet." *New York Times*, June 29, 2015.

CHAPTER 6: THE REVENGE OF RETAIL

US e-commerce statistics courtesy of US Census.

US farmers' market statistics courtesy of USDA.

Alter, Alexandra. "The Plot Twist: E-Book Sales Slip, and Print Is Far from Dead." *New York Times*, September 22, 2015.

Bell, David R., Jeonghye Choi, and Leonard Lodish. "What Matters Most in Internet Retailing." *MIT Sloan Management Review*, September 18, 2012.

Bloom, Ari. "In a Digital World, Physical Retail Matters More Than Ever." *Business of Fashion*, March 4, 2014.

Bonanos, Christopher. "The Strand's Stand: How It Keeps Going in the Age of Amazon." *Vulture*, November 23, 2014.

Chapman, Matthew. "Foyles and Waterstones Reap Rewards of Print Resurgence as Online Growth Slows." *Retail Week*, January 7, 2015.

Cima, Rosie. "Why the Comic Book Store Just Won't Die." *Priceonomics*, May 5, 2015.

Currid-Halkett, Elizabeth. "What People Buy Where." *New York Times*, December 13, 2014.

D'Onfro, Jillian. "Four Years Ago Gilt Groupe Was the Hottest Startup in New York—Here's What Happened." *Business Insider*, February 21, 2015.

Dorf, David. "Pure-Play Retail Is Doomed." *Oracle Commerce Anywhere Blog*, March 12, 2015.

"The Four Horsemen," talk by Scott Galloway at DLD15, available on YouTube, https://www.youtube.com/watch?v=XCvwCcEP74Q.

Gibson, Megan. "E-books Go Out of Fashion as Book Sales Revive." *Time*, January 9, 2015.

Griffith, Erin. "Counterpoint: Groupon Is Not a Success." *Fortune*, March 20, 2015.

——. "Fab Was Never a Billion-Dollar Company." *Fortune*, January 22, 2015.

Gustafson, Krystina. "Millennials Don't Want to Shop Where You May Think." CNBC, May 28, 2014.

Halkias, Maria. "Supermarkets Consider Replacing Self-Checkout Lanes." *Dallas Morning News*, July 7, 2011.

Heyman, Stephen. "Assessing the Health of Independent Bookshops." *New York Times*, February 25, 2015.

"Independent Bookstores Are on the Rise Despite Digital Competition." *Michigan Radio*, March 10, 2015.

Lacy, Sarah. "Andreessen Predicts the Death of Traditional Retail. Yes: Absolute Death." *Pando*, January 30, 2013.

McCrum, Robert. "Whisper It Quietly, the Book Is Back . . . and Here's the Man Leading the Revival." *The Guardian*, December 14, 2014.

Osnos, Peter. "How 'Indie' Bookstores Survived (and Thrived)." *Atlantic*, December 2, 2013.

Rigby, Darrell. "E-Commerce Is Not Eating Retail." *Harvard Business Review*, August 14, 2014.

———. "Online Shopping Isn't as Profitable as You Think." *Harvard Business Review*, August 21, 2014.

"The Rise of the Independent Bookstore." *Huffington Post Books*, May 29, 2015.

Ruiz, Rebecca. "Catalogs, After Years of Decline, Are Revamped for Changing Times." *New York Times*, January 25, 2015.

Salmon, Kurt. "The Store Strikes Back." KurtSalmon.com, March 8, 2013.

Schwartz, Barry. *The Paradox of Choice: Why More Is Less.* Harper, 2005.

Streitfeld, David. "Selling E-Commerce While Avoiding Amazon." *New York Times*, June 5, 2015.

———. "To Gain the Upper Hand, Amazon Disrupts Itself." *New York Times*, December 1, 2014.

Thau, Barbara. "Beware, Retailers: Ignore Millennials at Your Own Risk." *Forbes*, October 10, 2014.

Underhill, Paco. *Why We Buy: The Science of Shopping.* Simon & Schuster, 1999.

Valloppillil, Sindhya. "Why Consumer-Facing E-Commerce Is Broken." *Business Insider*, April 28, 2013.

Wahba, Phil. "Barnes & Noble's Stores Provide Relief as Online Sales Plunge." *Fortune*, March 3, 2016.

CHAPTER 7: THE REVENGE OF WORK

Autor, David H. "Polanyi's Paradox and the Shape of Employment Growth." Abstract, MIT, NBER, and JPAL, August 11, 2014.

Bender, Morgan, Benedict Evans, and Scot Kupor. "U.S. Technology Funding—What's Going On?" Andreessen Horowitz presentation, June 2015.

Brynjolfsson, Erik, and Andrew McAfee. *Race Against the Machine: How the Digital Revolution Is Accelerating Innovation, Driving Productivity, and Irreversibly Transforming Employment and the Economy*. Digital Frontier Press, 2012.

———. *The Second Machine Age: Work, Progress, and Prosperity in a Time of Brilliant Technologies*. W. W. Norton & Co., 2016.

———. "Why Workers Are Losing the War Against Machines." *Atlantic*, October 26, 2011.

Brynjolfsson, Erik, Andrew McAfee, and Michael Spence. "New World Order." *Foreign Affairs*, July/August 2014.

Caramanica, Jon. "The Next Branding of Detroit." *New York Times*, August 21, 2013.

Carr, Nicholas. *The Glass Cage: Automation and Us*. W. W. Norton & Co., 2014.

Crawford, Matthew. *Shop Class as Soul Craft: An Inquiry into the Value of Work*. Penguin Press, 2009.

Davidson, Adam. "Don't Mock the Artisanal-Pickle Makers." *New York Times Magazine*, February 15, 2012.

Ford, Martin. *Rise of the Robots: Technology and the Threat of a Jobless Future*. Basic Books, 2015.

Krugman, Paul. "The Big Meh." *New York Times*, May 25, 2015.

Lanier, Jaron. *You Are Not a Gadget*. Thorndike, 2010.

LeDuff, Charlie. *Detroit: An American Autopsy*. Penguin, 2013.

Maraniss, David. *Once in a Great City: A Detroit Story*. Simon & Schuster, 2015.

McNeal, Marguerite. "Rise of the Machines: The Future Has Lots of Robots, Few Jobs for Humans." *Wired*, April 2015.

Miller, Claire. "As Robots Grow Smarter, American Workers Struggle to Keep Up." *New York Times*, December 15, 2014.

Mirani, Leo. "The Secret to the Uber Economy Is Wealth Inequality." *Quartz*, December 16, 2014.

Moy, Jon. "On Shinola, Detroit's Misguided White Knight." *Four Pins*, March 26, 2014.

Nocera, Joe. "Is Motown Getting Its Groove Back?" *New York Times*, June 2, 2015.

Raffaelli, Ryan. "Mechanisms of Technology Re-emergence and Identity Change in a Mature Field: Swiss Watchmaking, 1970–2008." *HBS Working Knowledge*, December 12, 2013.

Rushkoff, Douglas. *Program or Be Programmed: Ten Commands for a Digital Age.* Soft Skull Press, 2011.

Spence, Michael. "Labor's Digital Displacement." Council on Foreign Relations, May 22, 2014.

Trudell, Craig, Yuki Hagiwara, and Ma Jie. "Humans Replacing Robots Herald Toyota's Vision of Future." *Bloomberg*, April 7, 2014.

Williams, Alex. "Shinola Takes Its 'Detroit Cool' Message on the Road." *New York Times*, January 6, 2016.

CHAPTER 8: THE REVENGE OF SCHOOL

Barshay, Jill. "Why a New Jersey School District Decided Giving Laptops to Students Is a Terrible Idea." *Hechinger Report*, July 29, 2014.

Blume, Howard. "L.A. School District Demands iPad Refund from Apple." *Los Angeles Times*, April 16, 2015.

Boyd, Danah. "Are We Training Our Students to Be Robots?" *Bright*, April 7, 2015.

Brenneman, Ross. "Before Buying Technology, Asking 'Why?'" *EdWeek*, June 18, 2014.

Carr, Nicholas. "The Crisis in Higher Education." *Technology Review*, September 27, 2012.

———. *The Shallows: What the Internet Is Doing to Our Brains.* W. W. Norton & Co., 2010.

Catalano, Frank. "Tech Happens: When Tablets and Schools Don't Mix." *GeekWire*, October 9, 2013.

Chiong, Cynthia, Jinny Ree, Lori Takeuchi, and Ingrid Erickson. "Comparing Parent-Child Co-Reading On Print, Basic, and Enhanced e-Book Platforms." Cooney Center, Spring 2012.

Colby, Laura. "News Corp.'s $1 Billion Plan to Overhaul Education Is Riddled with Failures." *Bloomberg Businessweek*, April 7, 2015.

Cordes, Colleen, and Edward Miller. "Fool's Gold: A Critical Look at Computers in Childhood." Alliance for Childhood, 2000.

DeAmicis, Carmel. "A Q&A with 'Godfather of MOOCs' Sebastian Thrun After He Disavowed His Godchild." *Pando*, May 12, 2014.

Dodd, Tim. "UNE Shuts Down Its Loss-Making MOOCs." *Financial Review*, August 25, 2014.

Edmundson, Mark. "The Trouble with Online Education." *New York Times*, July 19, 2012.

Emma, Caitlin. "Finland's Low-Tech Take on Education." *Politico*, May 27, 2014.

Helfand, Duke. "Reading Program Didn't Boost Skills." *Los Angeles Times*, February 7, 2005.

Hembrooke, Helene, and Geri Gay. "The Laptop and the Lecture: The Effects of Multitasking in Learning Environments." *Journal of Computing in Higher Education*, Fall 2003.

Herold, Benjamin. "After Ed-Tech Meltdown, a District Rebounds." *Ed-Week*, January 27, 2015.

Holstead, Carol. "The Benefits of No-Tech Note Taking." *Chronicle of Higher Education*, March 4, 2015.

Kachel, Debra. "School Libraries Are Under Attack." *New Republic*, July 13, 2015.

Kamenetz, Anya. "The Inside Story on LA Schools' iPad Rollout: 'A Colossal Disaster.'" *Hechinger Report*, September 30, 2013.

Konnikova, Maria. "Will MOOCs Be Flukes?" *New Yorker*, November 7, 2014.

Lewin, Tamar. "After Setbacks, Online Courses Are Rethought." *New York Times*, December 10, 2013.

Lewin, Tamar, and John Markoff. "California to Give Web Courses a Big Trial." *New York Times*, January 15, 2013.

McNeish, Joanne, Mary Foster, Anthony Francescucci, and Bettina West. "Exploring e-Book Adopters' Resistance to Giving Up Paper." *International Journal of the Book*, 2014.

———. "The Surprising Foil to Online Education: Why Students Won't Give Up Paper Textbooks." *Journal for Advancement of Marketing Education*, Fall 2012.

McNeish, Joanne E. and Barbara Kolan. "Confronting the Illusion of Technological Expertise Among College and University Students." Ted Rogers School of Management, Ryerson University, and Achva Academic College, 2014.

———. "A Cross-Cultural Study on Digital Delivery of Academic Course Content." Ted Rogers School of Management, Ryerson University, and Achva Academic College, 2014.

Miller, Larry, Bethany Gross, and Robin Lake. "Is Personalized Learning Meeting Its Productivity Promise?" CRPE, May 2014.

Miron, Gary, and Jessica Urschel. "Understanding and Improving Full-Time Virtual Schools." National Education Policy Center, July 2012.

Oppenheimer, Todd. *The Flickering Mind: Saving Education from the False Promise of Technology.* Random House, 2004.

Powers, William. *Hamlet's BlackBerry: A Practical Philosophy for Building a Good Life in the Digital Age.* Harper, 2010.

Rich, Motoko. "Kindergartens Ringing the Bell for Play Inside the Class-room." *New York Times,* June 9, 2015.

Rockmore, Dan. "The Case for Banning Laptops in the Classroom." *New Yorker,* June 6, 2014.

Sana, Faria, Tina Weston, and Nicholas J. Cepeda. "Laptop Multitasking Hinders Classroom Learning for Both Users and Nearby Peers." *Computers & Education,* October 2012.

Schuman, Rebecca. "The King of MOOCs Abdicates the Throne." *Slate,* November 19, 2013.

Shirky, Clay. "Why I Just Asked My Students to Put Their Laptops Away." *Medium,* September 8, 2014.

Strauss, Valerie. "Too Much Tech? An Argument for Keeping Schools Low-Tech." *Washington Post,* August 26, 2014.

"Students, Computers and Learning." OECD Publishing, 2015.

Vigdor, Jacob L., and Helen F. Ladd. "Scaling the Digital Divide: Home Computer Technology and Student Achievement." *Urban Institute,* June 2010.

Warschauer, Mark, and Morgan Ames. "Can One Laptop per Child Save the World's Poor?" *Journal of International Affairs,* Fall/Winter 2010.

Zakaria, Fareed. "Why America's Obsession with STEM Education Is Dangerous." *Washington Post,* March 26, 2015.

CHAPTER 9: THE REVENGE OF ANALOG, IN DIGITAL

Bezos, Jeff. Interview on the *Charlie Rose Show,* November 15, 2012.

Bilton, Nick. "Steve Jobs Was a Low-Tech Parent." *New York Times,* September 10, 2014.

Clarke, Peter. "When Did Analog Steal Digital's Mojo?" *Electrical Engineering Time,* May 28, 2015.

Danzig, Richard. "Surviving on a Diet of Poisoned Fruit Reducing the National Security Risks of America's Cyber Dependencies." Center for New American Security, July 2014.

Evans-Pughe, Christine. "Photonic Computers Promise Energy-Efficient Supercomputers." *Engineering and Technology Magazine,* December 15, 2014.

Honan, Matt. "This Is Twitter's Top Secret Project Lightning." *BuzzFeed News,* June 18, 2015.

Kelly, Kevin. *Cool Tools: A Catalog of Possibilities.* kk.org, 2014.

———. *What Technology Wants.* Viking Press, 2010.

Lanks, Belinda. "Evernote Has More Office Supplies to Sell." *Bloomberg Businessweek,* August 26, 2014.

Lohr, Steve. "If Algorithms Know All, How Much Should Humans Help?" *New York Times*, April 6, 2015.

McMillan, Robert. "Darpa Has Seen the Future of Computing . . . and It's Analog." *Wired*, August 22, 2012.

Shachtman, Noah. "In Silicon Valley, Meditation Is No Fad. It Could Make Your Career." *Wired*, June 18, 2013.

Wagner, Kurt. "There's a Shiny New Trend in Social Media: Actual Human Editors." *re/code*, June 24, 2015.

EPILOGUE: THE REVENGE OF SUMMER

Bisby, Adam. "Roam Free: The Case for Digital Detox at Camps." *The Globe and Mail*, June 25, 2015.

Brody, Jane E. "Screen Addiction Is Taking a Toll on Children." *New York Times*, July 6, 2015.

Heffernan, Virginia. "Magic and Loss." *New York Times Magazine*, February 18, 2011.

Holson, Laura M. "The IRL Social Clubs." *New York Times*, October 1, 2014.

Keim, Brandon. "Screens May Be Terrible for You, and Now We Know Why." *Wired*, March 18, 2015.

Pinker, Susan. *The Village Effect: How Face-to-Face Contact Can Make Us Healthier and Happier*. Random House, 2014.

Turkle, Sherry. *Alone Together: Why We Expect More from Technology and Less from Each Other*. Basic Books, 2011.

———. *Reclaiming Conversation: The Power of Talk in a Digital Age*. Penguin Press, 2015.

Index

David Sax is a journalist specializing in business and culture. His writing appears regularly in *Bloomberg Businessweek* and *The New Yorker's Currency* blog. He is the author of two books, including *The Tastemakers* and *Save the Deli*, which won a James Beard Award for Writing and Literature. He lives in Toronto.

PublicAffairs is a publishing house founded in 1997. It is a tribute to the standards, values, and flair of three persons who have served as mentors to countless reporters, writers, editors, and book people of all kinds, including me.

I. F. STONE, proprietor of *I. F. Stone's Weekly*, combined a commitment to the First Amendment with entrepreneurial zeal and reporting skill and became one of the great independent journalists in American history. At the age of eighty, Izzy published *The Trial of Socrates*, which was a national bestseller. He wrote the book after he taught himself ancient Greek.

BENJAMIN C. BRADLEE was for nearly thirty years the charismatic editorial leader of *The Washington Post*. It was Ben who gave the *Post* the range and courage to pursue such historic issues as Watergate. He supported his reporters with a tenacity that made them fearless and it is no accident that so many became authors of influential, best-selling books.

ROBERT L. BERNSTEIN, the chief executive of Random House for more than a quarter century, guided one of the nation's premier publishing houses. Bob was personally responsible for many books of political dissent and argument that challenged tyranny around the globe. He is also the founder and longtime chair of Human Rights Watch, one of the most respected human rights organizations in the world.

.   .   .

For fifty years, the banner of Public Affairs Press was carried by its owner Morris B. Schnapper, who published Gandhi, Nasser, Toynbee, Truman, and about 1,500 other authors. In 1983, Schnapper was described by *The Washington Post* as "a redoubtable gadfly." His legacy will endure in the books to come.

Peter Osnos, *Founder*